Impacts of Human Population on Wildlife

A British Perspective

Wildlife and the countryside are highly valued by people in the UK, and for good reason. Healthy habitats are invaluable assets and promote human well-being. However, they are under increasing threat from, among other things, relentless urban expansion and intensive modern agriculture. These pressures stem largely from a major underlying cause – the high and growing population of humans living in the UK. This book provides an overview of wildlife in the UK and its recent status, factors contributing to wildlife declines, trends in human numbers, international deliberations about the impacts of human population growth and the implications for the future of wildlife conservation in the UK. The evidence-based text includes comparisons of wildlife declines and their causes in other countries, providing a global perspective. This book is for ecologists, naturalists and conservation biologists studying and working in academia or in consultancies, as well as all those interested in wildlife conservation.

TREVOR J. C. BEEBEE is Emeritus Professor of evolution, behaviour and environment at the University of Sussex, UK, where he taught from 1976 to 2012. He has published over 200 papers, articles and books, including *Climate Change and British Wildlife* (Bloomsbury, 2018) which received the Marsh Award from the British Ecological Society for the most influential ecology book of 2019. He is also a trustee of Amphibian and Reptile Conservation, and a Fellow of the British Herpetological Society and the British Naturalists Association.

ECOLOGY, BIODIVERSITY AND CONSERVATION

The world's biological diversity faces unprecedented threats. The urgent challenge facing the concerned biologist is to understand ecological processes well enough to maintain their functioning in the face of the pressures resulting from human population growth. Those concerned with the conservation of biodiversity and with restoration also need to be acquainted with the political, social, historical, economic and legal frameworks within which ecological and conservation practice must be developed. The new Ecology, Biodiversity and Conservation series will present balanced, comprehensive, up-to-date and critical reviews of selected topics within the sciences of ecology and conservation biology, both botanical and zoological, and both 'pure' and 'applied'. It is aimed at advanced final-year undergraduates, graduate students, researchers and university teachers, as well as ecologists and conservationists in industry, government and the voluntary sectors. The series encompasses a wide range of approaches and scales (spatial, temporal and taxonomic), including quantitative, theoretical, population, community, ecosystem, landscape, historical, experimental, behavioural and evolutionary studies. The emphasis is on science related to the real world of plants and animals rather than on purely theoretical abstractions and mathematical models. Books in this series will, wherever possible, consider issues from a broad perspective. Some books will challenge existing paradigms and present new ecological concepts, empirical or theoretical models, and testable hypotheses. Other books will explore new approaches and present syntheses on topics of ecological importance.

Ecology and Control of Introduced Plants
Judith H. Myers and Dawn Bazely

Invertebrate Conservation and Agricultural Ecosystems
T. R. New

Risks and Decisions for Conservation and Environmental Management
Mark Burgman

Ecology of Populations
Esa Ranta, Per Lundberg, and Veijo Kaitala

Nonequilibrium Ecology
Klaus Rohde

The Ecology of Phytoplankton
C. S. Reynolds

Systematic Conservation Planning
Chris Margules and Sahotra Sarkar

Large-Scale Landscape Experiments: Lessons from Tumut
David B. Lindenmayer

Assessing the Conservation Value of Freshwaters: An International Perspective
Philip J. Boon and Catherine M. Pringle

Insect Species Conservation
T. R. New

Bird Conservation and Agriculture
Jeremy D. Wilson, Andrew D. Evans, and Philip V. Grice

Cave Biology: Life in Darkness
Aldemaro Romero

Biodiversity in Environmental Assessment: Enhancing Ecosystem Services for Human Well-Being
Roel Slootweg, Asha Rajvanshi, Vinod B. Mathur, and Arend Kolhoff

Mapping Species Distributions: Spatial Inference and Prediction
Janet Franklin

Decline and Recovery of the Island Fox: A Case Study for Population Recovery
Timothy J. Coonan, Catherin A. Schwemm, and David K. Garcelon

Ecosystem Functioning
Kurt Jax

Spatio-Temporal Heterogeneity: Concepts and Analyses
Pierre R. L. Dutilleul

Parasites in Ecological Communities: From Interactions to Ecosystems
Melanie J. Hatcher and Alison M. Dunn

Zoo Conservation Biology
John E. Fa, Stephan M. Funk, and Donnamarie O'Connell

Marine Protected Areas: A Multidisciplinary Approach
Joachim Claudet

Biodiversity in Dead Wood
Jogeir N. Stokland, Juha Siitonen, and Bengt Gunnar Jonsson

Landslide Ecology
Lawrence R. Walker and Aaron B. Shiels

Nature's Wealth: The Economics of Ecosystem Services and Poverty
Pieter J. H. van Beukering, Elissaios Papyrakis, Jetske Bouma, and Roy Brouwer

Birds and Climate Change: Impacts and Conservation Responses
James W. Pearce-Higgins and Rhys E. Green

Marine Ecosystems: Human Impacts on Biodiversity, Functioning and Services
Tasman P. Crowe and Christopher L. J. Frid

Wood Ant Ecology and Conservation
Jenni A. Stockan and Elva J. H. Robinson

Detecting and Responding to Alien Plant Incursions
John R. Wilson, F. Dane Panetta and Cory Lindgren

Conserving Africa's Mega-Diversity in the Anthropocene: The Hluhluwe-iMfolozi Park Story
Joris P. G. M. Cromsigt, Sally Archibald and Norman Owen-Smith

National Park Science: A Century of Research in South Africa
Jane Carruthers

Plant Conservation Science and Practice: The Role of Botanic Gardens
Stephen Blackmore and Sara Oldfield

Habitat Suitability and Distribution Models: With Applications in R
Antoine Guisan, Wilfried Thuiller and Niklaus E. Zimmermann

Ecology and Conservation of Forest Birds
Grzegorz Mikusiński, Jean-Michel Roberge and Robert J. Fuller

Species Conservation: Lessons from Islands
Jamieson A. Copsey, Simon A. Black, Jim J. Groombridge and Carl G. Jones

Soil Fauna Assemblages: Global to Local Scales
Uffe N. Nielsen

Curious About Nature
Tim Burt and Des Thompson

Comparative Plant Succession Among Terrestrial Biomes of the World
Karel Prach and Lawrence R. Walker

Ecological-Economic Modelling for Biodiversity Conservation
Martin Drechsler

Freshwater Biodiversity: Status, Threats and Conservation
David Dudgeon

Joint Species Distribution Modelling: With Applications in R
Otso Ovaskainen and Nerea Abrego

Natural Resource Management Reimagined: Using the Systems Ecology Paradigm
Robert G. Woodmansee, John C. Moore, Dennis S. Ojima and Laurie Richards

The Species–Area Relationship: Theory and Application
Thomas J. Matthews, Kostas A. Triantis and Robert J. Whittaker

Ecosystem Collapse and Recovery
Adrian C. Newton

Animal Population Ecology: An Analytical Approach
T. Royama

Why Conserve Nature? Perspectives on Meanings and Motivations
Stephen Trudgill

Invading Ecological Networks
Cang Hui and David Richardson

Hunting Wildlife in the Tropics and Subtropics
Julia E. Fa, Stephan M. Funk and Robert Nasi

The Life, Extinction, and Rebreeding of Quagga Zebras
Peter Heywood

Impacts of Human Population on Wildlife

A British Perspective

TREVOR J. C. BEEBEE

University of Sussex

CAMBRIDGE
UNIVERSITY PRESS

CAMBRIDGE
UNIVERSITY PRESS

University Printing House, Cambridge CB2 8BS, United Kingdom

One Liberty Plaza, 20th Floor, New York, NY 10006, USA

477 Williamstown Road, Port Melbourne, VIC 3207, Australia

314–321, 3rd Floor, Plot 3, Splendor Forum, Jasola District Centre, New Delhi – 110025, India

103 Penang Road, #05-06/07, Visioncrest Commercial, Singapore 238467

Cambridge University Press is part of the University of Cambridge.

It furthers the University's mission by disseminating knowledge in the pursuit of education, learning, and research at the highest international levels of excellence.

www.cambridge.org
Information on this title: www.cambridge.org/9781108833554
DOI: 10.1017/9781108985260

First published 2022

Printed in the United Kingdom by TJ Books Limited, Padstow Cornwall

A catalogue record for this publication is available from the British Library.

Library of Congress Cataloging-in-Publication Data
Names: Beebee, Trevor J. C. (Trevor John Clark), author.
Title: Impacts of human population on wildlife: a British perspective / Trevor J.C. Beebee, University of Sussex.
Description: First edition. | Cambridge, United Kingdom; New York, NY, USA: Cambridge University Press, 2022. | Series: Ecology, biodiversity and conservation | Includes bibliographical references and index.
Identifiers: LCCN 2022003213 (print) | LCCN 2022003214 (ebook) | ISBN 9781108833554 (hardback) | ISBN 9781108984782 (paperback) | ISBN 9781108985260 (epub)
Subjects: LCSH: Habitat conservation–Great Britain. | Great Britain–Population–Environmental aspects. | Nature–Effect of human beings on. | Wildlife conservation–Great Britain. | BISAC: NATURE / Ecology
Classification: LCC QH76.5.G7 B44 2022 (print) | LCC QH76.5.G7 (ebook) | DDC 333.95/40941–dc23/eng/20220225
LC record available at https://lccn.loc.gov/2022003213
LC ebook record available at https://lccn.loc.gov/2022003214

ISBN 978-1-108-83355-4 Hardback
ISBN 978-1-108-98478-2 Paperback

For Andrew and Caroline, and generations that follow

Contents

Preface

'We need the tonic of wilderness – to wade sometimes in marshes where the bittern and meadow-hen lurk, and hear the booming of the snipe.' These longings of naturalist Henry Thoreau in the wild woods of New England strike a chord nearly two centuries after his self-imposed exile at Walden Pond. He would be less happy there today, as the urine from swarms of swimmers has decimated the lake's fish populations. People have become a problem for wildlife, not just in New England but equally in old England and virtually everywhere else around the world. Yet the issue of human numbers is rarely discussed in any context, despite the fact that humans as much as any other species are at risk from ongoing population growth.

For me, the issue of people pressure imposed itself almost as soon as I became entranced by all the plants, animals and fungi that constitute the natural world around us. Spring is a time of anticipation, a regular reminder of our fascination with all things natural. Although especially wondrous for naturalists, few people remain untouched by the lengthening days and promise of a coming summer. And so it had been for me, increasingly intoxicated as a youth by wild places and the amazing wildlife there for the watching in my local countryside. But one day was different. It was April 1963, and hovering above the pond an intimidating machine was moving vast quantities of earth, infilling a treasure trove overflowing with water plants, dragonflies, beetles, frogs, newts and so much more. There had been no warning, at least none that had reached the ears of a schoolboy unversed in the ways of the world. This small pool had been my inspiration, awakening my first fascination with natural history. Over the previous 5 years I had become familiar with every soggy corner and to this day I could draw you a map of it. A shady bay beneath a hawthorn bush where, one autumn, a dozen newly emerged great diving beetles lay immobile on the silt; springs with frog choruses, masses of spawn, migrating toads and newts flitting around in the shallows; and warm summer days with hawker dragonflies patrolling tirelessly around the shores. Within a matter of hours, all this was gone, paving the way for a new housing estate. My adolescent brain made the obvious connection: more people means fewer wild places and less wildlife.

This apparently naïve interpretation of cause and effect has stuck with me despite the plethora of specific causes put forward to account for the ongoing degradation of Britain's environment. It seems astonishing that an obvious problem – too many people competing for space in a small country – has not only been largely overlooked but positively excluded from serious discussion at all levels of society. It was not always so. In the early 1970s, there was great interest in the impact of overpopulation, largely driven by Paul Ehrlich's 1968 book, *The Population Bomb*. Even the late but not much lamented US President Richard Nixon took the issue seriously. Sadly, and at great environmental cost, interest in overpopulation subsequently waned and we are now paying the price for decades of needless neglect.

In this book, I make the case that reasons usually proposed to account for the progressive degradation of our countryside are real enough but that the main ones are secondary to a single, primary cause: the attempt to accommodate more people than Britain can sustain without ongoing environmental damage. This is not a new idea, simply one that has not received the attention it deserves. Perhaps it soon will. In *Man Swarm, and the Killing of Wildlife* (Foreman, 2011) the subject of population pressures on wildlife is explored in some depth, but with a strong emphasis on the situation in North America. The issue has also been put forcefully by Shragg (2015). In *Move Upstream*, she makes a compelling case that virtually all of the problems identified by environmentalists as causes of concern are secondary to the consequences of overpopulation, defined as 'an undesirable condition where the number of existing human populations exceeds the carrying capacity of Earth'. Overpopulation, this emotive word, when used at all, has mainly been associated with developing countries. For some reason, problems from people numbers are considered relevant in Africa and Asia but largely ignored in Europe, despite the fact that in parts of this continent, including the UK, the impact of overpopulation on the environment is equally or more acute than in the developing world. This is particularly true in England, the most densely inhabited state of the union.

In the ensuing chapters, I consider first of all some assessments of the state of wildlife in Britain and how that has changed for the worse in recent decades. Proximal causes are then discussed, including direct persecution, physical developments of the kind that destroyed my childhood pond, invasive activities such as disturbance by increased footfalls, agricultural 'improvements' (which turn out to be the main offenders), climate change and the spread of new pathogens. 'Moving upstream' involves linking many of these well-known damaging activities to human numbers. Then comes the context of facts about how far and fast Britain's human population has increased, perceptions of the role of population size in human affairs, and how conservation measures might compensate for the ever-increasing population pressure.

Making the case for overpopulation problems in Britain is inherently more difficult than, for example, deducing the impact of climate change on wildlife. In the latter instance, there are many scientific papers demonstrating unequivocal links between climate warming and the distributions of plants and animals. Not so for population, a subject with precious few scientific studies to underpin analysis of its effects, particularly in Britain. The reason is simple to discern. The effects of overpopulation are mostly indirect, and connections have to be made between them and the proximal factors acting on wildlife mentioned above. Evidence must nevertheless be robust to be convincing, and there are certainly some relevant scientific papers, as well as a variety of specialist reports. Other aspects also differ from climate change. Computer modelling and experimental approaches are respectively more difficult or impossible in the context of overpopulation. Last but by no means least, some consequences of climate change have been beneficial to wildlife, at least in Britain. This is not true of overpopulation, where the impact has been overwhelmingly negative.

Failure to address the consequences of an escalating human population after initial interest in the subject waned some 40 years ago has had ramifications far beyond those involving wildlife. Instead of addressing the issue head on, healthy environments have been sacrificed to sustain ever-increasing numbers of people around the world. But the unhappy fate of dwindling wilderness is surely a bellwether for *Homo sapiens*. Thus far, we have postponed a day of reckoning as global resources have been squandered on a massive scale, but without a major rethink about population growth such a day will surely come.

In this book, I have included Latin names for species where first cited, but not in subsequent referrals. I have also used Britain and the UK interchangeably to minimise repetition in the text, realising of course that the two terms are not truly synonymous.

Acknowledgements

My thanks are due to Paul Ehrlich and George Monbiot for providing quotes for citation in this book, and also to David Richardson, Alistair Currie and John Guillebaud for providing information about the history and activities of the Population Matters charity. Dominic Lewis, Aleksandra Serocka, Jane Hoyle and Niranjana Harikrishnan of Cambridge University Press provided excellent support during the writing process, and Adriana de Palma of the Natural History Museum kindly provided national Biodiversity Intactness Index estimates. Ben Andrew of the Royal Society for the Protection of Birds, Arnold Cooke, Alistair Currie of Population Matters, Hans Dekker of the Saxifraga Foundation Jim Foster, Jane Gazzard of Plantlife, Chris Madden, Colin Munro of Colin Munro Photography, the British Geological Survey, Tim Newcomb of Newcomb Studios, Daniel Rosengren, Anju Sarpal and Bryony Davison of The Amphibian and Reptile Conservation Trust, Paul Sterry of Nature Photographers Ltd, Mike Toms of the British Trust for Ornithology, Charles Tyler of Exeter University and James Williams of the Joint Nature Conservancy Council generously permitted use of their images in book figures. Last but by no mean least, I am grateful to Maggie Beebee for meticulous proofreading as the chapters emerged.

Abbreviations

AD	ash dieback
AIDS	acquired immune deficiency syndrome
AONB	area of outstanding natural beauty
ASSI	Area of Special Scientific Interest (Northern Ireland)
BI	Biodiversity Intactness Index
BP	before present
BPS	basic payment scheme
bTB	bovine tuberculosis
BTO	British Trust for Ornithology
CAP	Common Agricultural Policy
CBD	Convention on Biological Diversity
Civitas	Institute for the Study of Civil Society
CO_2	carbon dioxide
COVID-19	coronavirus 2019
CPRE	Council for the Protection of Rural England
DDT	dichlorodiphenyltrichloroethane
DED	Dutch elm disease
DEFRA	Department for Environment, Farming and Rural Affairs
ELMS	Environmental Land Management Scheme
EU	European Union
FERA	Food and Environment Research Agency
FoE	Friends of the Earth
GDP	gross domestic product
GP	general practitioner
ICPD	International Conference on Population and Development
IPBES	Intergovernmental Science-Policy Platform on Biodiversity and Ecosystem Services
IUCN	International Union for the Conservation of Nature
IUD	intrauterine device
NBSAP	National Biodiversity Strategies and Action Plan

NFU	National Farmers' Union
NGO	non-governmental organisation
NHS	National Health Service
NNR	National Nature Reserve
NO_2	nitrogen dioxide
NPP	net primary production
ONS	Office for National Statistics
PCB	polychlorinated biphenyl
RHDV	rabbit haemorrhagic disease virus
RHS	Royal Horticultural Society
RSPB	Royal Society for the Protection of Birds
SAC	Special Area of Conservation
SDG	Sustainable Development Goal
SO_2	sulfur dioxide
SPA	Special Protection Area
SPS	single payment scheme
SSSI	Site of Special Scientific Interest
STC	Save the Children
UN	United Nations
UNFCCC	United Nations Framework Convention on Climate Change
WHO	World Health Organisation
WWF	World Wide Fund for Nature (formerly World Wildlife Fund)

1 · *Population Matters*

1.1 Problems with People

In the wilder parts of the UK, it is still possible to wander for many kilometres and not meet another living soul, but such places are increasingly rare and more or less non-existent in lowland England. The central theme of this book relates to that simple observation, focusing on human numbers and how they impact upon wildlife in Britain, and on other aspects of life in the UK and elsewhere. This first chapter's title matches that of a charity, Population Matters (https://populationmatters.org/), which has lobbied long and hard for the consequences of overpopulation to be taken seriously, and a gruelling task it has proved to be. Yet there are signs, discussed in later chapters, that awareness of population pressures is growing as the damaging effects of so many humans on the planet are increasingly hard to ignore. Whether this recognition leads to action designed to address population issues remains to be seen.

The evolutionary success of hominid apes would surely have defied the predictions of any computer model. Pathetically weak in tooth and claw, our early ancestors nevertheless managed to spread from the cauldron of their African birthplace across most of the Old World more than a million years ago. Small groups of hunter-gatherers used the products of burgeoning brain power to overcome a lack of inbuilt weaponry, and for untold aeons ranged over landscapes largely unmoved by their wanderings. There may have been fewer than 20,000 of these people breeding all over their range in those early days, a number smaller than current estimates of endangered apes such as chimpanzees (*Pan troglodytes*) today. But this was only the opening act of a play usurped by the products of another African genesis, this time of modern humans (*Homo sapiens*) some quarter of a million years past. With bigger brains and a correspondingly more sophisticated armoury, the newcomers followed their antecedents across Asia and beyond. By 15,000 years ago, every continent except Antarctica was home to *Homo sapiens*, while

Figure 1.1. Mammoth under attack by humans. *Source*: Wikipedia under CC Attribution-Share Alike 4.0 licence

the pioneering hominids that preceded them vanished in their wake. This new wave was altogether different from what had gone before. Although the evidence is debatable, it looks very much as if the increasing numbers of modern humans had a much greater impact on the environment than their predecessors. Not only did archaic human species disappear coincident with the increasing populations of modern man, but so did a range of megafauna including mammoths (*Mammuthus* spp.; Figure 1.1), mastodons (*Mammut* spp.) and the woolly rhinoceros (*Coelodonta antiquitatis*). Some of these animals were certainly victims of human predation, demonstrated by fossilised skeletons with associated spear points and damage from attack (Lister, 2014). Because many of these events coincided with dramatic climate changes through the Pleistocene glaciations, cause and effect can be difficult to establish but it looks very much as if human predation played a significant role in species exterminations from early in our history. And all this was long before human number escalated to multimillions, let alone the billions of us around today. We started as we meant to go on.

1.2 About This Book

It is all too clear that wildlife is declining in both species diversity and abundance across the globe, on a scale that looks increasingly like the start of a sixth mass extinction comparable with five others that ravaged the planet over the past 4 billion years. It is also indisputable that most if not all of the current declines are due, one way or another, to human activities. That a bipedal ape of no fixed abode could generate such overwhelming misery is astonishing and certainly without precedent. An important question is the extent to which this ongoing disaster relates directly to the numbers of people on the ground or in more complex ways to multifarious human activities. Conservation policies and actions have almost invariably focused on the latter issue, especially the ever-increasing consumption of global resources. Some prominent environmental campaigners such as George Monbiot have strongly emphasised the need to reduce resource exploitation and played down the role of an increasing human population:

> It is true that, in some parts of the world, population growth is a major driver of particular kinds of ecological damage, such as the expansion of small-scale agriculture into rainforests, the bushmeat trade and local pressure on water and land for housing. But its global impact is much smaller than many people claim. The formula for calculating people's environmental footprint is simple, but widely misunderstood: Impact = Population × Affluence × Technology (I = PAT). The global rate of consumption growth, before the pandemic, was 3% a year. Population growth is 1%. Some people assume this means that the rise in population bears one-third of the responsibility for increased consumption. But population growth is overwhelmingly concentrated among the world's poorest people, who have scarcely any A or T to multiply their P. The extra resource use and greenhouse gas emissions caused by a rising human population are a tiny fraction of the impact of consumption growth.

By contrast, Paul Ehrlich, the early and eminent promoter of policies for population reduction in the 1970s, is unrepentant:

> The human population has grown so large that roughly 40% of the Earth's land surface is now farmed to feed people – and none too well at that. Largely due to persistent problems with distribution, almost 800 million people go to bed hungry, and between one and two billion suffer from malnutrition. As a consequence of its booming population, *Homo sapiens* has taken much of the most fertile land to grow plants for its own consumption. But guess what? That cropland is generally not rich in food plants suitable for the caterpillars of the

15,000 butterfly species with which we share the planet. Few butterflies require the wheat, corn or rice on which humans largely depend. From the viewpoint of most of the Earth's wildlife, farming can be viewed as 'habitat destruction'. And, unsurprisingly, few species of wildlife have evolved to live on highways, or in strip malls, office buildings, kitchens or sewers – unless you count Norway rats, house mice, European starlings and German roaches. Virtually everything humanity constructs provides an example of habitat destruction.

From the point of view of this book, the fact that population growth is currently greatest in poor countries is not strictly relevant. What matters is that Britain has experienced a substantial growth rate for a long time, resulting in an extraordinarily high density of humans, particularly in England. Population growth and resource consumption are both problematic not just for the future of wildlife but also for that of our own species. However, in my view, the burgeoning human population is by far the most important driver of environmental damage in Britain. One way of addressing this controversy is to ask what exactly are the mechanisms by which humans damage wildlife in the UK, and to what extent would they be ameliorated if there were fewer of us around?

1.3 What Are the Issues?

People in Britain, as elsewhere, have ambivalent views about wildlife. For many, especially naturalists, there is unqualified pleasure in the countryside's plant and animal variety. For some, though, the opposite can be true. Farmers, for example, regularly wage an uncompromising war against arable weeds and insect pests. In Chapter 2, these relationships are explored, and the costs and benefits of wildlife in the UK are accounted. There is then an assessment of the current state of wildlife in our island archipelago, including documentation of the many ongoing species declines and of legislation put in place in attempts to protect them.

Three subsequent chapters outline the main human activities that impact adversely on wildlife, in increasing order of likely significance and where appropriate also in relation to Britain's increasing human population. Chapter 3 considers human activities that kill wildlife directly, including predation and collection. Evidently this kind of assault was in place millennia ago when Neolithic hunters took their toll of large Pleistocene mammals, but it has continued long after that. In relatively recent times, many species around the world have been exterminated or severely depleted by humans using increasingly effective armaments to

Figure 1.2. Bison skull pile. *Source*: https://en.wikipedia.org/wiki/File:Bison_skull_pile-restored.jpg; unknown author, taken in 1892

bring down their prey. Examples are legion. Dodos (*Raphus cucullatus*) in Mauritius, thylacines (*Thylacinus cynocephalus*) in Tasmania, Steller's sea cow (*Hydrodamalis gigas*) and all too many others have gone down as a result of overkill. A striking example of how effective firearms can be was the almost total annihilation of bison (*Bison bison*) on the North American plains. It has been estimated that some 50 million of these extraordinarily abundant animals were shot by white settlers in the nineteenth century (Figure 1.2), depriving native peoples of a staple food and leaving only a few hundred of them alive. Fortunately, the species has recovered substantially since this nadir, but the example shows how astonishingly devastating direct predation by humans can be.

Animals are not the only organism imperilled by direct persecution. Plant diversity is declining fast everywhere, and although this is mostly due to habitat change, some species are also at risk from collection or overharvesting. The Himalayan may apple (*Sinopodophyllum hexandrum*), which produces a chemotherapeutic chemical (podophyllotoxin) useful in cancer treatments, has suffered from overharvesting as well as habitat loss and is currently classified as endangered. The question is to what extent has direct persecution caused wildlife declines in Britain, and how does it relate to the numbers of people still engaged in doing it?

Arguably the most obvious signs of human impact on the countryside relate to increases in infrastructure, notably the ongoing expansion of urban sprawl together with road and railway networks across the countryside. These causes of wildlife declines are discussed in Chapter 4. When the Nature Conservancy was created in the aftermath of the Second World War, ever more building over our green and pleasant land was perceived to be by far the greatest danger to our native plants and animals. Of course, human infrastructure is not a new phenomenon. However, mediaeval and earlier dwellings, with trackways interconnecting them, were in most places too sparse to impact significantly on the landscape. Indeed, Roman roads may have inadvertently facilitated the spread of so-called archaeophyte plants, including some of our most attractive wildflowers, brought over by the invaders from mainland Europe. This innocuous situation was set to change by the start of the eighteenth century, since when increasing numbers of people have created much larger settlements –ultimately great cities – with networks of busy roads betw.een them. The industrial revolution added bustling factories, the combined results of all this being a reduction in land space for wildlife, effective fragmentation of previously interconnected habitats, and an onslaught of air and water pollution spreading way beyond the infrastructure itself. Nor is damage restricted to terrestrial and freshwater ecosystems. Offshore pollution has also been extensive. And there have been other undesirable consequences of these developments, including disturbance of sensitive species by increased background noise and by the footfall of more people walking, sailing and swimming in previously quiet places. Once again, the question becomes how much of the UK's declining biodiversity is due to these causes and to their relationship with an increasing human population?

What the newly established Nature Conservancy failed to anticipate were the ramifications of a nascent agricultural revolution, just getting underway in the mid-twentieth century with the advent of artificial fertilisers and pesticides. Omission of this risk assessment was understandable. The wider countryside was surely a haven for wildlife, more so than ever following the farming depression of the 1930s, and this benign situation was expected to continue. Chapter 5 describes what actually happened. Alarmed by a U-boat offensive that substantially reduced Britain's food imports during the war, it was judged politically imperative to prevent any future recurrence of such a threat. An attempt began to make the UK self-sufficient in food, and to that end farmers would receive government subsidies to industrialise the countryside with heavy machinery, large quantities of fertilisers, and the newly available herbicides and insecticides. Over subsequent decades, this intensification of farming has continued apace, and wildlife has paid the price for it. Impacts are not only evident at the level of hostile fields and declining hedgerows. Aquaculture with offshore fish farms require overharvesting of food species, while dredging has canalised rivers and wreaked havoc among sensitive marine communities on sea beds. Problems in other parts of the countryside started even earlier in the twentieth century. This time, in response to timber shortages during the First World War, it was decided to greatly increase plantation woodland across the country. Serried ranks of conifers sprang up, often in places with rare species unable to cope with such a dramatically changed habitat. Here, too, wildlife declines inevitably followed. Altogether, this range of major changes, extending over large swathes of the UK, has had the most serious consequences for our wildlife heritage fuelled by the dependence of increasing numbers of people on countryside products.

Not all threats to British wildlife are home-grown. Chapter 6 documents the consequences of three relatively new developments, climate change, an upsurge in pathogenic diseases and increases in disturbance. A warming world is witnessing substantial alterations in plant and animal populations, especially across the Arctic where temperatures are rising most steeply. Polar bears (*Ursus maritimus*) and walruses (*Odobenus rosmarus*) are increasingly vulnerable to decreases in the extent of summer sea ice, but wildlife in warm countries is also feeling the effects of greater heat. Extensive wildfires in eastern Australia during 2019–2020 are reckoned to have killed 3 billion animals, mostly reptiles, and several species may have been completely exterminated. The inhabitants of California, unlike some of their compatriots in Washington, need no convincing that climate change is real (Figure 1.3) as every year more of their houses go up in flames. In the UK climate change has resulted in milder winters, new species arriving from warmer climes, and declines in Arctic–alpine plants and some coastal seabird colonies. Wildlife diseases are also on the march, facilitated by increasing international trade and air travel. Among the most devastating of these is the pathogenic fungus *Batrachochytrium dendrobatidis*, a native of eastern Asia that has spread across the globe with devastating effects on many amphibians, some of which it has totally wiped out. Novel diseases such as this have arrived in

Figure 1.3. Wildfires raging in California. *Source*: Kent Nishimura/Los Angeles Times/Getty Images (A black and white version of this figure will appear in some formats. For the colour version, please refer to the plate section.)

the UK in recent decades, and here too there have been severe consequences for several species of plants and animals. The human population size in Britain is not directly linked to these developments, but connections with human numbers at a global scale can certainly be made.

1.4 About People Numbers

The final four chapters focus on various aspects of the human population size. That this has been growing exponentially across the world in recent decades is common knowledge. In late 2020, the estimated global total exceeded 7.8 billion people, with several extra mouths to feed being added every second. The pattern in Britain has broadly mirrored that in the rest of the world, and recent trends in the constituent countries of the UK are outlined in Chapter 7. The changes have been dramatic, with more than twice as many people living in Britain in 2020 than were present as recently as the Second World War. This doubling has happened within a single human lifetime. Previous and contemporary predictions of population growth are discussed with special consideration of how accurate, or otherwise, they have proved to be. Changes in population size are driven by three main factors: birth rate, death rate and levels of immigration/emigration. The relative importance of these drivers dictating events in the UK are considered, together with an assessment of how these factors have changed over time. Finally, there is a comparison of the human population size and extent of wildlife declines in Britain with the same features in some other countries. To what extent do such comparisons confirm or refute major impacts of human numbers on wildlife?

Chapter 8 is about people's perceptions of the UK population size. Impressions of the general public on this topic have been obtained via opinion polls, sampled at various times over the years. More selective assessments have investigated the views of particular factions within society. Many naturalists and scientists have long considered overpopulation as a matter of concern, while others, including conservation-based non-governmental organisations (NGOs), have often taken a more cautious approach. Various religious organisations have decried attempts by individuals to control family size, while economists have been generally hostile towards any limitation of population growth because of its demographic consequences: an increased proportion of elderly people would need support from a relatively smaller workforce. Politicians of all parties have mostly shied away from the topic altogether. Yet impacts on wildlife are only a small part of the price paid for high human numbers in Britain. Pressures on housing, road and airport expansion continue to increase with concomitant exacerbations of traffic jams, air pollution and overstretched public services. Is there an increased realisation that human population increase is not indefinitely sustainable in the UK?

Human population trends are increasingly of international concern. Several countries have implemented attempts at population control, the best known of which is China's one-child policy. The successes, failures and human rights issues of these efforts are considered in Chapter 9. International conventions, declarations and frameworks of varying relevance to population growth include the 2015 Paris Agreement on climate change and a Global Assessment Report on Biodiversity and Ecosystem Services by the Intergovernmental Science-Policy Platform. International summit meetings in Cairo and more recently in Nairobi specifically addressed population issues and made recommendations about future action to tackle associated concerns. To what degree have these exercises translated into meaningful action in the face of the continuing increase in human numbers around the world?

Finally, Chapter 10 provides an overview of population increases in relation to the specific causes of wildlife decline in the UK discussed in Chapters 3–6, as well as a comparison with reasons for wildlife declines in other countries. How much commonality is there? To what extent can the international and British biodiversity crises be related to human overpopulation as a primary cause? What humane strategies are available for controlling population growth? In the immediate and long-term futures, how can conservationists best improve the prospects for British wildlife in the face of increasing human numbers?

1.5 The UK as an Example of Population Pressure

In 2020, the UK had the 14th highest population density in the world, and the third highest in Europe after the Netherlands and Belgium (https://statisticstimes.com/demographics/countries-by-population-density .php). However, most people in the UK live in England, which has the fifth highest density in the world (after Bangladesh, Taiwan, South Korea and Rwanda), and the highest within Europe (www.statista.com/statis tics/281322/population-density-in-the-united-kingdom-uk-by-coun try/). Scotland, Northern Ireland and Wales are home to inspiring plants and animals in spectacular countryside vistas, but it is in the warm lowlands of England that most of the UK's biodiversity resides. Some examples of population densities in a range of countries are shown in Table 1.1.

Population density is a more appropriate indicator of environmental impact than population size because in some countries with big populations, people are spread over very large areas. Thus, the USA has a population of 330,000,000, which is about five times that of the UK but ranges across 9,000,000 km^2 of land compared with just 250,000 km^2 in the UK. Human encroachment on wildlife habitats is commonly associated with East African savannahs, but, as shown in Table 1.1, the population densities of Kenya and Tanzania are only a quarter of that in the UK as a whole and perhaps 16 per cent of that in England. Habitats and ecosystems in the tropics are very different from those in Europe, and it is comparisons within that densely peopled continent that are most relevant to investigating impacts of human numbers in Britain. Community ecologists have long recognised the importance of population densities in the dynamic processes of competition and predation. Wildflower diversity in meadowlands is quickly diminished by heavy competition from grass encroachment unless this is constrained by

Table 1.1. *Examples of human population densities in a range of countries around the world*

Country	No. of people per km^2 in 2020
Bangladesh	1175
South Korea	516
Netherlands	421
India	416
Belgium	377
China	377
Japan	338
UK	280
England	**430**
Scotland	70
Wales	151
Northern Ireland	136
Pakistan	275
Germany	233
Nigeria	217
Italy	200
Denmark	135
France	123
Portugal	111
Austria	106
Spain	93
Kenya	82
Ireland	70
Tanzania	59
USA	34
Brazil	25
Sweden	23
Norway	17
Canada	4
Australia	3
'World' average	**58**

suitable management (Peterken, 2013). Predation and/or competition from an increasing population of badgers (*Meles meles*) is associated with drastic declines in hedgehogs (*Erinaceus europaeus*) in England and Wales, although this species interaction is certainly not the only factor involved in hedgehog declines (Williams et al., 2018). Humans constitute another species with rapidly increasing numbers and the potential to precipitate damage to biodiversity. Subsequent chapters therefore focus on the main causes of wildlife declines in Britain and how they relate to population pressures there and in other European countries. Even so, invoking human numbers as the root cause of major wildlife declines should be treated as a hypothesis and not as a given fact. As science pioneer Francis Bacon warned as long ago as 1620:

> The human understanding, once it has adopted an opinion, collects any instances that confirm it, and though the contrary instances may be more numerous and more weighty, it does not notice them or else rejects them, in order that this opinion will remain unshaken.

This is all too true in so many walks of life, as political and environmental lobbyists still find to this day.

2 · *The State of British Wildlife*

2.1 Wildlife in the UK

Time and again we hear about the wildlife extravaganzas of tropical rainforests, the astonishing species diversities in the Earth's seas and oceans, and the bizarrely adapted plants and animals of deserts and Arctic wastes. Yet for a naturalist there can be no better place to call home than the UK. More than 70,000 species of plants, animals and fungi dwell in these sceptred isles (Hayhow et al., 2016), and few of even the most enthusiastic observers can hope to encounter more than a fraction of them in a lifetime. We have this startling estimate for the UK's biodiversity on account of the country's remarkably long and enviable history of wildlife recording. Even so, like distances to the stars, it is not the kind of number to which people can easily relate, and it gives a false impression of static, unchanging bounty. The reality is both more complex and more interesting. Species have been arriving and occasionally departing the British Isles ever since the last glaciers retreated some 10,000 years ago. Early human invaders, weaving cautiously across the marshes of now-submerged Doggerland, were hunter-gatherers reclaiming a home from which their ancestors had been driven by intolerable cold. This time people had come to stay but in an environment that was far from stable, and many other recolonisers have enjoyed a less secure future. Multiple species arrivals and extinctions followed the retreating ice, and not all of the newcomers turned up on their own accord. Almost half of the vascular plants in the British countryside today arrived within the last 200 or so years, propagated deliberately or accidentally by Victorian collectors and enthusiastic gardeners. Sycamores (*Acer pseudoplatanus*) and *Rhododendron* spp., rabbits (*Oryctolagus cuniculus*) and rats (*Rattus rattus* and *Rattus norvegicus*), among many other species, arrived after William of Normandy's last successful human invasion of these islands. The end result is an extraordinary landscape celebrated by poets, artists and naturalists for generations past. England's green fields and sombre woods, the startling mountain vistas of Scotland and Wales, the glorious coastline of Northern Ireland and the multiplicity of tumbling streams, awesome lakes and meandering rivers are an inspiration to anyone who steps outside their front door. Few other countries of similar size can boast such scenic variety, and it is this mixed mosaic that has allowed so many species of wildlife to prosper in the astonishing range of habitats that the UK has to offer. But all is not well with our wild world. The state of the British countryside at the start of the twenty-first century is not a happy one, and ever more people are becoming concerned about it. Before focusing on this plight, it is worth considering why wildlife has proved fascinating for so many people over the years, and what costs and benefits accrue from taking care of it.

2.2 The Naturalist's Tale

Boundless Joy

For those lucky enough to retain the excitement of visiting wild places beyond the curiosity of early childhood, the pleasures of wildlife encounters need no explanation. Changing seasons herald changing moods: melancholy as nights draw in, nagging anticipation as winter's stark vistas begin to falter and exaltation in the warmth of seemingly endless summer days. These are common emotions in all walks of life but are hugely enhanced for naturalists tuned in to wildlife that resonates with cycles of birth and death. Birch catkins suddenly unveiled on a damp and drizzly morning announce that spring is no longer on hold. Woodland's green mantle of fresh, sparkling foliage changes subtly but irrevocably into the darker hues of high summer. As if from nowhere, autumn dew drenches gorse and bramble and is adorned with sparkling curtains, courtesy of ubiquitous web spiders. And before we know it, the ground is crunching underfoot with frost and an icy skin spreads over ponds that, months earlier, teemed with irrepressible life. These rhythms play out predictably on Earth's orbital loop, but every year unveils its own distinctive signature to confuse our simple-minded expectations. Will it be an early year for frog spawn? How will the breeding season for visiting summer birds turn out? The uncertainty surrounding these annual delights ensures that natural history is eternally fascinating to those caught up in its passions. Everyone has particular favourites. For me, they are mostly water-related. Arrival of the first newts in our garden ponds has been a diary entry for more than 40 years. One night, nothing to see; then a glimpse of movement among the water weeds confirms that a smooth newt (*Lissotriton vulgaris*), probably a male (no crest yet), has made the trek from its winter refuge. And for all naturalists there are memorable firsts, as firmly fixed in the mind as a wedding day or the births of children: a great silver beetle (*Hydrophilus piceus*) creeping out of a mass of globular algae hauled out of a marshy ditch, or natterjack toads (*Epidalea calamita*) calling and spawning in sunlit sand dune pools. There is just so much to celebrate.

A History of Natural History

Finding delight in natural history has long been a tradition in Britain, perpetuated in prose and poetry from mediaeval to modern times. Francis Bacon, writing at the turn of the seventeenth century, premeditated the age of enlightenment with his exposition of the scientific method. One man's inspiration heralded a revolution that swirls around us with undiminished ferocity to this day, but he also noted that moss grows on the north side of trees, strange fungi spring up in the forests and glow-worms shine in the night (Eiseley, 1961). Bacon's extraordinary foresight has resonated through centuries of scientific advances, including an anticipation of their potential costs to a brave new world: 'Force maketh nature more violent in the returne' was a call for humility, anticipating the awesome power of science while its method was still in the cradle. The message should be remembered in an age when unsustainable exploitation of the Earth's resources is coming home to roost.

Bacon combined curiosity and astute observation in his encounters with the natural world. Many other naturalists followed suit. Essex-born John Ray was the first explorer of Britain's wildlife to leave detailed written accounts, especially of his beloved plants, a century after Bacon's time. Among other achievements, Ray was the first to attempt the definition of a species, an issue that remains contentious to this day. Botany was his first love, and *Historia Plantarum* (Ray, 1686, first volume) became a milestone in the story of Britain's natural history. Onwards to another century, and Gilbert White, pottering in his Selborne garden, is renowned for astute observations in and around this Hampshire village (White, 1789). By now, natural history was becoming an important element of scientific discovery. In correspondence with fellow enthusiasts such as Thomas Pennant, the author of *British Zoology* (another ground-breaking work) in 1776, White revealed both his passion for wildlife and a desire to understand what he saw around him. Where did swallows (*Hirundo rustica*) go in winter? Was the discovery of a water vole (*Arvicola amphibious*) nest in a chalky field far from water a rare event or a common but unrealised feature of this rodent's lifestyle? Voyagers to distant lands began to include naturalists among their crew, documenting new discoveries of animals and plants from faraway places. Joseph Banks endured the trials of James Cooke's first venture into the Pacific and returned with accounts of unimaginably diverse wildlife from far beyond Britain's shores. But it was another, later traveller who transformed an abiding interest into an intellectual revolution. The consequences of Charles Darwin's voyage on HMS *Beagle*, culminating in a theory of evolution based on natural selection, need no embellishment here. Yet again we have someone passionate about wildlife, but in this case also with a rare intellectual frisson able to transform our understanding of how nature works. Feeding this febrile mind was a simple sense of wonder that any naturalist would recognise. Darwin's gardener once remarked to a visitor that 'the poor man, he just stands and stares at a yellow flower for minutes at a time. He would be better off with something to do' (Eiseley, 1979). I think not.

The blossoming of natural history in Victorian England was not an exclusively male preserve. Many women, all too often unsung, were every bit as entranced by the mysteries of wild nature. Mary Anning, an early explorer of Dorset's fossil-rich coast, is considered one of the 10 most influential women in science. Her finds of ancient reptiles in particular are prominent in museums across the land. Eleanor Ormerod's interests included 'injurious insects' and innovative, sometimes dangerous experiments. Why are some amphibians avoided by predators that could easily catch and devour them? Knowing the animal to be toxic, she pressed part of the back and tail of a live great crested newt (*Triturus cristatus*) between her teeth (Gadow, 1909):

> The first effect was a bitter astringent feeling in the mouth, with irritation of the upper part of the throat, numbing of the teeth more immediately holding the animal, and in about a minute from the first touch of the newt a strong flow of saliva. This was accompanied by much foam and violent spasmodic action, approaching convulsions, but entirely confined to the mouth itself. The experiment was immediately followed by headache lasting for some hours, general discomfort of the system, and half an hour after by slight shivering fits.

Such an eccentric, devil-may-care approach to nature study would surely generate apoplexy among health and safety enforcers today.

Natural history in nineteenth-century Britain had a very different ethos from what was to come later. Vast numbers of non-native plants and animals, collected by voyagers returning from abroad, were released or escaped into the British countryside. Most perished but some prospered, and the damaging legacy of a few, such as grey squirrels (*Sciurus carolinensis*), is still with us. It was also a time of increasing persecution by gamekeepers of 'vermin', mostly birds of prey and predatory mammals, and of a passion for collecting reflected today in trays of butterflies and birds' eggs still gathering dust in museum drawers. With the benefit of hindsight, Victoriana could be judged a period of thoughtless exploitation of wildlife, but this was long before any need for conservation became evident. And it wasn't all bad. Those museum collections remain a valuable resource for researchers trying to rediscover the ecology of past times, and a spin-off from the interest in all things wild was the emergence of companies producing equipment for the field naturalist. Nets, traps, weed drags, microscopes, vivaria and aquaria became available for the first time and at reasonable cost. The simple pond net remains a vital companion to those of us that delight in chasing the inhabitants of ponds and ditches. I could barely function without mine.

Nature Writing

Gilbert White's legacy goes beyond factual accounts and perceptive observations. His letters convey much more than that, literary masterpieces that hold our attention centuries after the celebrated parson put pen to paper. 'Hen-harriers fly low over heaths or fields of corn, and beat the ground regularly like a pointer or setting-dog' – one sentence chosen at random on a page of *The Natural History and Antiquities of Selborne* (White, 1789). Other talented writers have followed White's tradition, delighting in the profound beauty of the British countryside and its wildlife. Richard Jefferies left us, after a tragically short life, with memorable accounts of Wiltshire's fields and woods in *Wildlife in a Southern County* (Jefferies, 1879). Agriculture in the late nineteenth century was utterly different from what we know today; hedges were thick and rampant, while fields were tended by exhausting manual labour. 'Those who labour in the fields seem to have far fewer holidays than the workers in towns.... But the farmer and the labourer work on much the same – the closing of banks and factories in no way interfering with the tilling of the earth or the tending of cattle.' Birds and insects abounded. William Henry Hudson, an Argentinian by birth, spent much of his adult life travelling around southern England. He, too, is an inspiring read. Among many engaging accounts, *Nature in Downland* (Hudson, 1900) recalls summer visitors in Sussex villages: 'Alfriston and Ditchling appeared to be the most favoured by swallows and martins. . .and if a woman came out of her front door and emptied a bucket of dirty water in the road, half a dozen martins would quickly drop down on the wet place to get a little mortar for the nest they were building.' Alas, how rare a sight this has become in Sussex a century on.

Nature writing continues to enthral us and has experienced a revival in recent decades. Richard Mabey has been a gold standard for this genre since the 1970s, with a host of books and articles to his credit. In *A Brush with Nature* (Mabey, 2010) we are reminded of his contributions to the *BBC Wildlife* magazine, for me a highlight of that erstwhile publication. Now Mabey is one of many, and by the turn of the twenty-first century a veritable plethora of talented authors were on the scene, including an increasing number of female writers. Melissa Harrison's *Rain: Four Walks in English Weather* (2016) is a prime example, vividly bringing home the role of a wildly varying climate in our love of the outdoors. These accounts move us in ways that natural history books acutely focused on studies of habitats and species cannot. Unsurpassed companions on dark winter evenings, our thoughts are lifted to drift and dream, all thanks to a few skilled wordsmiths.

Academics and Amateurs

The turn of the twentieth century witnessed the birth of professional natural history, embodied by the employment in academic institutions of people dedicated to investigating wild plants and animals in their natural state as well as in laboratories. The science of ecology was born. For those experiencing the competitive edge of recent university life, a culture in which money and prestige seem to dictate almost everything, the halcyon days of early twentieth-century academia are unimaginable. The new ecologists were passionate naturalists free to pursue whatever took their fancy with few bureaucratic or financial constraints. E. B. 'Henry' Ford chased butterflies with his father in sunlit meadows and later combined ecology with the 'new' genetics to understand variation in those enchanting insects (Ford, 1945). Eccentricity was no barrier to success; Ford used the space beneath his desk for thinking, and allegedly ended his lectures exactly on the hour even if this was in mid-sentence. Frank Balfour-Browne (1962) obtained a lectureship at Cambridge by 'catching the eye' of a professor, and went on to document the life histories and distributions of British water beetles. By mid-century, there was employment to be had in government agencies as well as universities. For three decades after the Second World War, naturalists seeking a salary to follow their dreams enjoyed a purple patch. Pioneers of ecology and conservation thrived at this time, far too many to list (but see Marren, 2005). Winifred Frost investigated the fishes of Windermere at the prestigious laboratories of the Freshwater Biological Association; Julian Huxley, grandson of 'Darwin's Bulldog' Thomas Henry Huxley, pioneered ethological studies of water birds and was a joint founder of the World Wildlife Fund (WWF; now the World Wide Fund for Nature); Norman Moore made a home in the Nature Conservancy's research station at Monks Wood, pursuing dragonflies and bringing to national attention the massive declines in England's lowland heaths; Derek Ratcliffe, an outstanding field naturalist, discovered the disastrously damaging effects of dichlorodiphenyltrichloroethane (DDT) on raptor egg shells (Figure 2.1), and became the Nature Conservancy's chief scientist; Miriam Rothschild was a code-breaker at Bletchley Park but loved to watch butterflies and went on to pioneer habitat restoration

Figure 2.1. Cartons of peregrine falcon eggshells, which were thinned by the birds' exposure to the pesticide DDT. *Source*: Galen Rowell/The Image Bank Unreleased/ Getty Images

using wildflower seeds; Jeremy Thomas's ground-breaking study of butterfly ecology resulted in the successful reintroduction of the large blue (*Phengaris arion*) to England; and Niko Tinbergen's fieldwork on animal behaviour won him the only Nobel prize thus far awarded to someone known primarily as an outstanding naturalist, although the prize was for Physiology and Medicine. Regrettably, ecology remains a poor relation in the scientific world and does not have its own Nobel category.

It was not to last. By the end of the 1980s, support for wildlife study and nature conservation by the public purse was waning and government research institutes were contracting or closing down altogether. Fortunately, charitable NGOs have risen to the challenge of engaging people in natural history and its conservation to a degree that now parallels or even outstrips government commitments, a sad indictment of recent political priorities. Academic work goes on, but increasingly sophisticated ecological studies published in learned journals are inaccessible to any but the most scientifically minded naturalists. There are, of course, exceptions to this trend. The impact of climate change on British wildlife has attracted the attention of scientists well able to communicate their findings to a wide audience. Tim Sparks, Chris Thomas (brother of Jeremy) and James Pearce-Higgins are among those alerting us to the effects of global warming on plants and animals in Britain (Beebee, 2018). But most of the increasing numbers of people interested in natural history engage with it as amateurs, following a part-time hobby and content to watch, listen and wonder at the marvels that surround us in the British countryside. Among them, however, are outstanding naturalists for whom description as an amateur verges on pejorative. John Clegg was a hugely accomplished limnologist, supporting his interest by working as curator of Haslemere and then Selborne museums. His book *The Observer's Book of Pond Life* (Clegg, 1956) was one of my youthful inspirations. Nobody knew the wildlife of East Anglia better than Ted Ellis, another museum man (this time in Norwich) before retiring to run his own nature reserve at Wheatfen Broad. Generous to a fault, Ted helped many aspiring naturalist find their way around the wild places of Norfolk and Suffolk. I was one such beneficiary. Then again, there was a time when the Nature Conservancy employed field staff wise in the ways of their home patch, largely free of the bureaucracy and financial scrimping that bedevils its successor (Natural England) today. Colin Tubbs was one of these, with unparalleled knowledge and a love of the New Forest, a jewel in our wilderness crown with its huge diversity of habitats and species. School teaching was another way of funding a wildlife hobby. Ernest Neal pioneered investigations into the lives of badgers while making a living at Taunton school. And the medical profession sometimes provided yet another way in. Malcolm Smith, a physician at the court of Siam, went on snake-hunting expeditions in his spare time. On retirement to England he wrote the first comprehensive book on British amphibians and reptiles (Smith, 1951; another of my inspirations), much of it based on his personal experiences in the field. Some of the most enthusiastic naturalists have found multifarious ways of supporting a passion for wildlife over and above what most folk have to be content with.

The Media

Post-war Britain was the setting for a major development in the appreciation of wildlife: an explosion of new books and magazines, including species monographs, habitat accounts and identification guides for all things living, from birds to mosses and liverworts. Wildlife books were hardly a new idea, and there were notable productions in Victorian times, but most were written in styles of little appeal to twentieth - or twenty-first-century readers. Thomas Bell's *A History of British Quadrupeds* (1837) was a comprehensive guide and a great improvement on what had gone before, but illustrations were limited to woodcuts and, somewhat esoterically, the text included dogs and sheep as well as wild animals. Newer methods of publication with high-quality colour printing generated more attractive works at reasonable cost, and the famous Collins *New Naturalist* series started the revolution with its first contribution, *Butterflies*, by E. B. Ford in 1945. Seventy years on, the series remains seminal, with more than 100 volumes to its credit. In bookshops today, there is something for every natural history interest; we are spoilt for choice and there is no sign of a slowdown in public interest. *British Wildlife* magazine set up shop in 1989 and goes from strength to strength, recently carving another series of well-illustrated wildlife books. *BBC Wildlife* magazine reaches many thousands of readers, as do wildlife television programmes that feature spellbindingly high photographic quality. By the start of the twenty-first century, media outlets were popularising wildlife in ever more imaginative ways and generating a host of new household names. David Attenborough and Peter Scott set the pace in the 1950s, and now we have, among others, Chris Packham, Iolo Williams and Lucy Cooke. Inevitably, the Internet too has had an increasing influence on natural history, with mobile phone photography allowing anyone with an interesting picture to identify a species online. Thousands of people use iSpot (www.ispotnature.org/) to find out what is in their picture and give the lie to any suggestion that new technology inevitably leads to less engagement with nature.

There is an important message underlying the long-standing interest in Britain's wildlife and wild places. Very large numbers of people are engaged. The Royal Society for the Protection of Birds (RSPB) alone has a steadily increasing membership of well over a million; county wildlife trusts are supported by tens of thousands, and a host of smaller specialist NGOs for bats, amphibians, reptiles, dragonflies, butterflies and so on attract hundreds or thousands of enthusiastic followers. More than 4.5 million people in the UK are currently signed up to one or more wildlife conservation organisations. Television programmes that major on British wildlife, such as *Springwatch* and *Countryfile*, regularly attract more than 2 million viewers and are among the most popular items on the broadcasting schedules. The combined total membership of all of the UK's main political parties adds up to less than that of the RSPB alone, but our democratic representatives manage to ignore wildlife issues for most of the time. One consequence, as we shall see, is that the pleasure provided by nature to so many people has been jeopardised as this magnificent heritage degrades before our eyes. However, before addressing that depressing situation, it is worth considering other aspects of wildlife's place in Britain in a materialistic way that most naturalists think little about.

2.3 Costs and Benefits

Costs

Judging by the membership of wildlife organisations, perhaps 10 per cent of people in the UK are positively engaged with the country's natural history. But what about the other 90 per cent? Polls suggest (see Chapter 8) that many of us are sympathetic to wildlife and its conservation, but for some people wild plants and animals create significant personal costs. These are rarely as damaging as the price paid in countries that still sustain serious predators. In Africa, more than 200 people are killed annually by lions (*Panthera leo*) alone. Very many more, there and around the world, die after attacks from a wide range of other animals including dangerous herbivores such as African buffalo (*Syncerus caffer*).

Comparable tragedies in Britain are rare but not unheard of. Collisions of deer with motor vehicles cause several hundred injuries and around 20 deaths every year in the UK, and the number is rising. Also on the increase is the incidence of Lyme disease, an unpleasant infection transferred to passing walkers by the bites of ticks (*Ixodes ricinus*) lurking on pathside vegetation. Nevertheless, British wildlife is generally benign with respect to personal danger or physical suffering.

Costs of wildlife in the UK are therefore mostly of a different kind. Every year, a combination of taxpayers and charitable or private funders spend substantial sums on wildlife in Britain, either in the cause of conservation or, more often, for control or eradication. Table 2.1 summarises the types of expenditure currently ongoing. The national statutory conservation agencies provide several hundred million taxpayer pounds annually, mostly to support the running of nature reserves and to fund various stewardship schemes promoting conservation in the wider countryside. Other government bodies also support conservation projects, notably the Environment Agencies and Forestry Services of England, Scotland and Northern Ireland (those for Wales are subsumed into Nature Resources Wales). Wildlife charities make major contributions to the same ends. Statutory and private landowners also pay out for more negative aims, notably controlling various types of damage consequent on wildlife activities and including mitigation measures for development of areas supporting protected species. Foremost among these costs, however, are the widespread applications of pesticides by farmers to control crop-damaging insect infestations and competitive weeds. There seem to be no readily available national financial statistics on this high-level activity with its major impacts on wildlife conservation, and the numbers shown in Table 2.1 are estimates derived from areas under cultivation and average costs per hectare of pesticide application. The figure given is therefore certain to be inaccurate but sufficient to give a realistic idea of the enormous financial burden for farmers still in thrall to the agrochemical industries. Other expenditures for wildlife control are individually smaller but add up to a substantial sum. In addition to the cost of deer traffic collisions mentioned above, which amounts to tens of millions of pounds each year, the Forestry Commission and the Department of the Environment, Farming and Rural Affairs (DEFRA) have provided the estimates shown in Table 2.1 for culling to reduce deer damage to woodlands (summarised by Lewis, 2021). Control of non-native species including Japanese knotweed (*Fallopia japonica*), *Rhododendron* spp., zebra mussels (*Dreissena polymorpha*), grey squirrels

Table 2.1. *Approximate annual costs of wildlife control in the UK*

Activity	Approximate costs	Comments
Conservation work		
Natural England	>£165 million	Financial support of UK statutory conservation agencies, funded by government. Information from the most recent public accounts.
Scottish Natural Heritage	£53 million	
Natural Resources Wales	>£160 million, but covers many activities other than nature conservation	
Council for Nature Conservation and the Countryside (Northern Ireland)	£20 million	
Environment agencies and forestry services	>£200 million	
Agroenvironment schemes:		All schemes subsidised by the European Union until 2020. Most basic payments are farm subsidies, with a minority supporting conservation management.
Basic payment scheme	£1.5 billion	
Countryside stewardship schemes	£150 million	
Wildlife charities:		Financial support provided by private memberships and grants.
RSPB	£100 million	
The Wildlife Trusts	£140 million	
Other charities (including Amphibian and Reptile Conservation, Bat Conservation, Buglife, Butterfly Conservation, Froglife, Mammal Society, Plantlife, Woodland Trust)	£10s of millions	
Coping with damage due to wildlife		
Mitigation schemes:		Efforts to translocate or support populations under threat from development.
Great crested newts	>£20 million	
Bats	Probably >£10 million	
Pesticide protection of crops	>£3 billion	Sourced from FERA/DEFRA data.

Table 2.1. (*cont.*)

Activity	Approximate costs	Comments
Damage from and/or control of non-native species including: rabbits, grey squirrels, house mice, brown rats, mink, rhododendron, Japanese knotweed, giant hogweed, New Zealand flatworm, signal crayfish, New Zealand stonecrop	>£1.7 billion	Sourced from DEFRA and Williams et al. (2010).
Damage from and/or control of native species:		
Deer	~£9 million	Damage to forestry (Forestry Commission Scotland) and crops (DEFRA).
Foxes	>£10 million	
Badgers	£3–6 million	Culling in bovine tuberculosis areas.
Pigeons and gulls	£830,000	Deterrence and shooting (pigeons).
Birds (miscellaneous)	Probably £10,000s	Licensed control of cormorants and buzzards. Starling raids on cattle feed.
Other pests (miscellaneous)	Probably £10,000s	Includes mole trapping, and cockroach, woodworm and fly infestations.

DEFRA, Department of the Environment, Farming and Rural Affairs; FERA, Food and Environment Research Agency.

and rabbits costs perhaps £1.7 billion per year (Williams et al., 2010). In most cases, the figures are approximate; no detailed UK statistics are available, as far as I can see, for the total annual costs of many other endeavours such as cormorant (*Phalacrocorax carbo*) or cockroach control. In these situations, the figures are based on the product of individual treatments and the likely number of them per year, in both cases usually with high levels of variation or uncertainty. Nor is the table comprehensive, although omissions of specific actions such as fencing to protect fisheries from otters (*Lutra lutra*), replacing crops lost to wild boar (*Sus scrofa*) and so on are thought to be minor components of the total spent each year.

Probably some £6 billion are spent each year on wildlife-related activities in Britain, the bulk of which (maybe 80 per cent) is devoted to damage limitation. Pesticide applications in agriculture constitute by far the largest cost, more than 50 per cent of the total expenditure, while conservation is served by around £1 billion per year, less than 20 per cent of UK spending on all wildlife issues. Comparative estimates for defence are £37 billion, education £20 billion, healthcare £150 billion and civil service running costs more than £11 billion. With an annual UK gross national product of around £2 trillion, the equivalent of only about 0.05 per cent of that is dedicated to conserving our wildlife heritage. This amounts to less than £25 per year per adult, the equivalent of two half-decent bottles of wine; hardly seems extravagant, does it?

Benefits

Sacrificing those wine bottles turns out to be a very small price to pay for the benefits that a healthy, biodiverse environment can bring. Sadly, politicians rarely comprehend let alone acknowledge the value of nature and have been increasingly reluctant to fund wildlife conservation from the public purse. The income of Natural England halved between 2010 and 2018, and this worthy organisation was one of the most hard-hit victims of the government drive for austerity and cut-backs in public spending. Such a short-sighted strategy, based on ravaging services that generate the least public outcry, is a recipe for ongoing environmental degradation. This attitude surely has to change. The value of wildlife conservation needs restating with sufficient emphasis to turn the tide of destruction outlined later in this chapter.

A good economic case for environmental protection is readily made. For those of us imbued with an unquestioning love of nature, resorting to financial argument seems like a necessary but tedious evil – akin, for me, to taking strenuous exercise in the hope that some good will come of it. That is probably a sound assumption, but I can still think of many more enjoyable ways to spend my time. One way of encouraging the protection of wildlife that might work, therefore, is to engage economists with evidence of financial or other tangible benefits that arise from our natural heritage. To do this convincingly might generate a broader level of support than naturalists alone can muster, and may counter a belief that conservation is an expensive luxury with no obvious benefits to society. This is no mean task. The economics community is not well known for its environmental sympathies, as evidenced by the reluctance of successive chancellors to take climate change seriously. It often looks easier to put a price on damage done by wildlife in Britain, as outlined in the above section, than to demonstrate financial rewards accruing from its preservation. Nevertheless, sound efforts have been made to show how wildlife works for us in terms of money rather than "just" aesthetic perceptions. One consequence of bringing accountants onto the scene has been the arrival of a new vocabulary over the past 30 years. We have been entreated to take on board, among other jargon, 'natural capital', 'green infrastructure' and, especially, 'ecosystem services'. What does all this amount to?

Natural capital is broadly defined as the UK's stock of natural resources (Office for National Statistics reports: www.ons.gov.uk/economy/environmentalaccounts/bulletins/uknaturalcapital/ecosystemserviceaccounts1997to2015). It includes non-renewable items such as fossil fuels and minerals, as well as renewables such as timber, fisheries, water sources and various ecosystem services, of which more later. According to the ONS, the total value of the UK's natural capital in 2018 was around £761 billion. In 2016, the exploitation of natural capital generated some £16 billion to the UK economy. However, between 2009 and 2014, it was estimated that the natural capital reserve declined by 32 per cent, largely due to the utilisation of non-renewables, especially fossil fuels. It is therefore not straightforward to make a connection between natural capital and any contribution within it that 'renewable' wildlife makes to the UK economy. With green infrastructure, a term used to describe the networks of natural and semi-natural features that provide benefits to people (https://ecosystemsknowledge.net/resources/tools-guidelines/green-infrastructure), we can do a little better. It usually involves local projects in which costs and benefits can be evaluated in greater detail than is normally possible with natural capital. Green infrastructure produces attractive landscapes in towns and cities, but accomplishes much more than that. In Glasgow, a green infrastructure project is based on an investment of £33 million per year over 35 years, a very substantial commitment. The outcome will include conversion of 3,500 ha of derelict land into greenspace; creation of more than 30 km of street trees, green roofs and ponds; 5,500 new parks, gardens, woodlands and meadows; 4,500 allotments; 500 km^2 of wildlife habitats; and more than 1,000 km of new travel routes (cycle paths and walkways). These are surely valuable improvements to the quality of life in and around Glasgow, but again there is no easy way to calculate any financial gains from them. In north-west England, however, there has been an attempt to quantify the positive outcome of green infrastructure projects. One estimate is a total benefit of £2.6 billon for the area including the creation of 100, 000 jobs in environmental and related fields. The National Forest in the region has attracted 330,000 visitors since 1995 and generated an annual income of £128 million. There are some interesting indirect effects too. Property values are on average 8 per cent higher near green spaces than in more urban settings, generating a total increase in property value of some £4.2 billion. These are impressive sums but still leave us some way distant from a clear accounting specifically of wildlife contributions.

The concept of ecosystem services is perhaps the closest we can get to putting a monetary value on renewable nature. Defined by DEFRA as 'services provided by the natural environment that benefit people', they constitute much of the renewable section of natural capital, and include a wide range of activities, as summarised in Table 2.2.

This seemingly comprehensive list omits, or perhaps subsumes, some specific activities that have particular relevance for wildlife and habitats. The value of ecosystem services has been investigated by the UK National Ecosystem Assessment service, from 2011 onwards (http://uknea.unep-wcmc.org/). Pollination is an increasingly important issue relevant to food production and is provided for free by the UK's diminishing insect populations. The value of this service to farmers has been estimated at £430 million per year. Fish are a marine wildlife food resource worth over £600 million annually. Trawlermen in the North Sea concentrate on succeeding in a hard and dangerous job, rarely appreciating that their harvest is considered as an ecosystem service and one that is under some threat, as described in Chapter 3. Flood control is an increasingly important activity that intact ecosystems can provide as climate change increases the frequency of water disasters. Flooding typically costs anything from £80 to £4,100 per ha to redress, and the role of wetlands in the provision of clean water contributes as much as £1.5 billion to the national economy. Flood plain nature reserves can hold huge quantities of water, benefitting otherwise at-risk areas by saving millions of pounds. Greenhouse-gas sequestration by forests and peatlands would, by artificial means, cost more than £500 million in total, and the amenity value of coasts and wetlands for tourism is worth at least £1.3 billion per year.

Table 2.2. *Ecosystem services recognised in the UK*

Service type	Examples
Provisioning	Crops, fruit, fish Fibre, fuel (timber, wood) Natural medicines Genetic resources Ornaments (flowers, bulbs)
Regulation	Air quality (effects of forests etc. removing air pollutants) Climate change (greenhouse-gas sequestration by forests and moors) Erosion (vegetation cover) Water purification (such as by reed beds) Bioremediation of waste (pollutant removal, storage, dilution, burial)
Cultural	Spiritual satisfaction Art inspiration Society adhesion (e.g. support of fishing communities) Aesthetic beauty Heritage (landscape and species preservation) Recreation and tourism attractions
Supporting	Soil formation/retention by appropriate farming methods Nutrient cycling, maintaining functional integrity of habitats Primary productivity, e.g. generation of organic material by photosynthesis in healthy forests Water cycling, sustaining provision and use of water resources Oxygen production, e.g. by sustaining extensive forest cover Habitat provision, by land acquisition and management to maintain biodiversity

A recently recognised benefit of attractive, wildlife-rich environments not explicit in Table 2.2 is the role they can play in the improvement of mental health (Figure 2.2). Mental ill-health is on the rise in the UK, and increasing efforts are underway to explore treatments that do not depend exclusively on the use of drugs. Patients suffering from depression and other mental disorders evidently gain great comfort from exposure to the UK's wondrous countryside (Bragg & Atkins, 2016), and it has been estimated that full use of this free treatment could save national healthcare providers more than £2 billion, essentially by reducing the need for health-centre beds and residential care homes. However, it is both difficult and arguably unnecessary to put an accurate price on such an important subject. The mental health benefits from 'green' exposure include a range of improvements, listed

Figure 2.2. Interacting with nature has benefits for mental health. *Source*: Oscar Wong/Moment/Getty Images

by Bragg & Atkins (2016) as: psychological restoration and increased general mental well-being; reduction in depression, anxiety and stress-related symptoms; improvement in dementia-related symptoms; improved self-esteem, confidence and mood; increased attentional capacity and cognition; improved happiness, satisfaction and quality of life; sense of peace, calm or relaxation; feelings of safety and security; increased social contact; inclusion and sense of belonging; and increase in work skills, meaningful activity and personal achievement. With an estimated one in four people experiencing a 'significant' mental health problem in any one year, the value to individuals and to society of alleviating these issues is surely alone sufficient to justify effective protection of the UK's glorious countryside and rich wildlife heritage.

Estimates of the monetary value of wildlife and wild places are inevitably incomplete, sometimes of dubious accuracy and make a range of uncertain assumptions. It is impossible to draw up comprehensive, accurate balance sheets of costs and benefits of wildlife to the UK economy because too much of the necessary information is either difficult to define precisely or is just not available. However, what there is suggests that wildlife and wildlife habitats are far from being a drain on UK finances. On the contrary, they make many important contributions, and the benefits probably outweigh the costs by a substantial margin. Producing a more robust comparison is a long-overdue job for the Treasury and should persuade even the most cocooned chancellor that more, not less investment in environmental protection will have positive economic impacts in future. For many naturalists and others out to enjoy the countryside, including me, these monetary comparisons are at best a distraction from the sheer beauty of the UK landscape and its plethora of plants and animals. But they have to be attempted, and may hopefully increase support for conserving what is for many people so self-evidently worthwhile.

2.4 The State of Nature Today

A Tale of Woe

That wildlife has declined substantially in Britain over recent decades is not news to any naturalist. For those old enough to remember the countryside of decades past, there is, all too often, a special sadness about youthful reflections on how things used to be. In the 1960s, house martins (*Delichon urbicum*), flighty harbingers of a coming summer (Figure. 2.3), arrived in droves at the Surrey village of my teenage years. A diary entry of 4 November 1962 reports on the last martins heading south that autumn, but now these engaging little birds are rare visitors to the region. History repeated itself at the University of Sussex, where abundant martin nests were considered a nuisance by unsympathetic groundsmen in the 1970s. They have no problem today – the summer flocks have long since forsaken their old homes. By the early 2010s, there were still small numbers of house martins around the Somerset village where we now live, and a couple of nests on our new home. What a grim story it is, this progressive disappearance of an enchanting symbol of summer joy. My experiences are not unusual. Martins have deserted Gilbert White's Selborne village in Hampshire where, 250 years ago, he described their habits and the excitement of their annual return. It is good to know that house martins are faring better in northern Britain, but England's debilitating experience is mirrored in much of northern Europe (Piersma, 2014). The likely primary cause, farming intensification, has much wider impacts than on this one species and is considered extensively in Chapter 5.

Personal experiences are vivid but ultimately subjective. Just as one swallow doesn't make a summer, losing one favourite animal does not of

Figure 2.3. House martins, one of many species declining in England. *Source*: Arterra/Universal Images Group/Getty Images (A black and white version of this figure will appear in some formats. For the colour version, please refer to the plate section.)

itself indicate an impending nationwide catastrophe. Unfortunately, in this case, the observations do have such overtones. Growing and widespread concern about declines of British wildlife came to a head in the first decade of the twenty-first century, leading the RSPB to coordinate a widespread investigation into what is going on. The UK has a long history of wildlife recording, and there is a plethora of both government organisations and NGOs with specialist interests in a wide range of taxonomic groups. Using information patiently acquired over decades, it proved possible to assess trends for a wide range of species since the 1970s. Among those contributing hard-won information were, in addition to the RSPB, the Centre for Ecology and Hydrology, the People's Trust for Endangered Species, Bat Conservation, Butterfly Conservation, Marinelife, Natural England, the National Trust, Amphibian and Reptile Conservation, Buglife, the Mammal Society, the British Trust for Ornithology (BTO), the Marine Biological Association, the Natural Environment Research Council, and the Botanical Society of Britain and Ireland. What a formidable array of evidence-generators this has amounted to, the outcome of which were three State of Nature reports (the most recent in 2019) and a robust scientific analysis of this remarkable and unprecedented investigation (Burns et al., 2018). The expectation is to repeat State of Nature reports on a trienniel cycle, with the next one due in 2022.

The story that emerged is a profoundly depressing one. Both in the relatively long term (1970–2013) and in a recent shorter term (2002–2019), more than half of the several thousand species for which data were available were in decline. In both time periods, over 40 per cent of these plants and animals suffered moderate or strong, not minor, decreases. There was no indication that declines had slackened or ceased in the most recent, short-term analysis compared with the longer one. Data for several thousand species represent less than 10 per cent of the tens of thousands that occur in the UK, but there is no reason to believe that the sample is unrepresentative of the total. The investigation made great efforts to ensure, as far as possible, that this was true, even though some groups, such as fungi, were relatively underrepresented while others, such as vertebrates, were in the opposite category. For some particularly well-studied species, it is possible to show the rate of decline over very long periods, and independent sources backed up the State of Nature conclusions. The BTO has accrued data for many British birds over several decades and has assessed trends in groups associated with particular habitats. Figure 2.4A shows the fate of four such groups, all of which have declined since 1970. Farmland species, in particular, have diminished dramatically. Similarly, insects are in big trouble, with many pollinators in serious decline, as shown in Figure 2.4B (https://hub.jncc.gov.uk/assets/3de3abe1-d7d1-417e-9684-1348dd8b9a5a). Reductions of 40 per cent or more of our moth species since the 1960s are commonplace in England, although fortunately these insects have fared better in Scotland (Conrad et al., 2004).

Although a majority of the plants and animals we should worry about today are unknown to most people and familiar only to those with specialist interests, many others are well known and much loved by the general public. Cornflowers (*Centaurea cyanus*), corncockles (*Agrostemma githago*), chicory (*Cichorium intybus*) and an array of other colourful wildflowers have disappeared, together with the hay meadows they once brought to life, over most of Britain. Since the Second World War, more than 97 per cent of these cornerstones of the countryside have gone under the plough. *The Wind in the Willows* brought 'Ratty' to life for countless young readers, and 30 years ago the 'plops' of panicking water voles disappearing beneath the surface could be heard around the banks

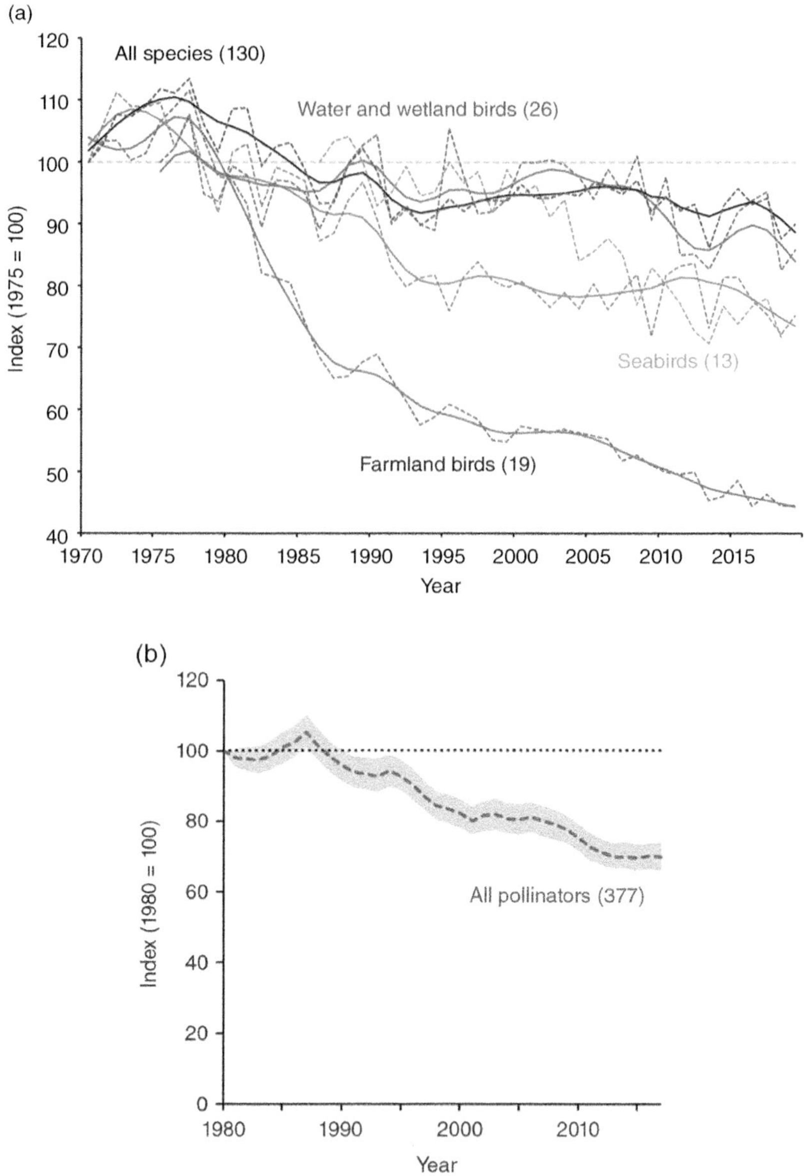

Figure 2.4 (a) Changes in the abundance of breeding birds of woodland, farmland, water and wetlands, and all species in the UK. *Source*: BTO, DEFRA, JNCC, RSPB. (b) Line graph showing indicator trend of 377 pollinator species (wild bees and hoverflies) declines in the UK, with shaded 90 per cent confidence limits. *Source*: Defra 2020. UK Biodiversity Indicators. Indicator D1c. Status of pollinating insects. Scheme; *Source*: Bees, Wasps & Ants Recording Society; Biological Records Centre (supported by UK Centre for Ecology & Hydrology and Joint Nature Conservation Committee); hoverfly recording (A black and white version of this figure will appear in some formats. For the colour version, please refer to the plate section.)

of ponds and ditches throughout the land. Not any more. This charming little rodent has declined by more than 90 per cent since the 1970s, as has another mammal even more familiar to many of us, in gardens as well as in the countryside. Not so long ago, almost every car journey of any length usually encountered one or more flattened hedgehog along the way. How long since you last saw one? Much more likely today is to mourn the passing of a deceased badger on the roadside, one of a minority of animals that has fared well in recent decades. Farmland birds have had a really bad time, with a long list of species on a downward slope. Yellowhammers (*Emberiza citrinella*), bullfinches (*Pyrrhula pyrrhula*), skylarks (*Alauda arvensis*) and plenty of others are much rarer now than they were in the not-too-distant past. The same goes for many summer migrants, including house martins, swifts (*Apus apus*), reed warblers (*Acrocephalus scirpaceus*) and cuckoos (*Cuculus canorus*). To hear that long-hallowed announcement of spring I must now resort to the Somerset Levels, where the cuckoos' call still resonates across the reed beds. There is no use listening in or around villages elsewhere in the county. And, beneath most peoples' radar, has been the insect disaster. Flies, bees, moths, butterflies and other vital invertebrate components of a healthy ecosystem have declined enormously since the 1960s. This catastrophe has verged on the subliminal for most people, but was brought sharply into focus by McCarthy (2015). Once upon a time, driving on a summer night ended with a windscreen opaque with the splattered bodies of myriads of moths. In most parts of Britain, the resultant cleaning problems are long gone. This insect armageddon attracted widespread press coverage after reports from Germany of flying insects declining by 75 per cent over the past 25 years even on nature reserves, but this should have come as no surprise. The Rothamsted Research centre has been documenting similar declines in Britain since the late 1960s.

In 1965, the International Union for the Conservation of Nature (IUCN) invented the Red List, in which the status of species all around the world was scientifically assessed and then described using a traffic-light system. Green species were OK while, moving towards the red end of the spectrum, categories became increasingly desperate through 'vulnerable', 'endangered' and ultimately 'critically endangered'. The idea was subsequently extended to individual countries, including the UK. Of 7,964 British species assessed by 2012 Red List criteria, 1,057 (13 per cent) were placed in the three categories considered to be at high risk, while 142 species (2 per cent) were already extinct. Sixty-seven out of

247 birds are on the endangered list. For some of these, such as curlews (*Numenius arquata*), the UK is home to more than a quarter of the global population. The Red List is regularly revisited and not all the news is bad. Raptors such as buzzards (*Buteo buteo*) and red kites (*Milvus milvus*), the latter almost extinct in the mid-twentieth century, are no longer considered to be endangered following dramatic recent recoveries. Unfortunately, downhill moves dominate the picture. Since 2009, 15 birds have been demoted towards extinction on the Red List. Charismatic members of the UK avifauna at risk of departing the islands altogether include nightingales (*Luscinia megarhynchos*) and turtle doves (*Streptopelia turtur*). Among the insects, many species are also shifting redwards. The high brown fritillary butterfly (*Argynnis adippe*) is seriously endangered, and several others, including the large copper (*Lycaena dispar*) and large tortoiseshell (*Nymphalis polychloros*), are already extinct or almost so. Most recently, a measure of 'biodiversity intactness' has been proposed as a key indicator of whether biodiversity has fallen below a threshold necessary to meet 'society's needs' including the adequate provision of ecosystem services. Most of Britain is now beneath this threshold; the UK registers as having a biodiversity intactness well below the global average, and is the 13th lowest out of 236 countries where comparable assessments have been made. This is not a score to be proud of. By a range of criteria, it is clear that UK wildlife is in dire straits and is now seriously impoverished relative to most of the world.

Why have the increasingly dire straits of wildlife in Britain received so little attention? Much of the bad news has appeared briefly on national bulletins and some of it has featured in parliamentary debates recorded in Hansard. Even so, most people remain largely unaware of the wildlife crisis in our countryside. Media reporting biases, probably unintentional, must be partly to blame for public ignorance about what is going on. Depressing news items such as international disasters, murders and the like usually dominate the reporting agenda. Occasionally, there is a fillip of something better, often proffered to counterbalance a predominantly dispiriting news output. This is where wildlife issues most often surface. We learn about the return of white-tailed eagles (*Haliaeetus albicilla*), great bustards (*Otis tarda*) and beavers (*Castor fiber*) or similar heart-warming success stories. The emphasis with wildlife reporting has, at least until recently, been the opposite of major news slants, with good news more frequently presented than bad. This view through rose-tinted binoculars easily leads to false impressions about how the majority of UK plants and animals are faring.

Why Is This Happening?

Although the main reasons for wildlife declines in the UK have been recognised by conservationists for many years, attempts to stem the tide have had very limited success, as demonstrated by the dour assessments described above. Legislators have not been entirely blind to the predicament of wildlife in Britain's countryside, and several acts of parliament since the 1950s, foremost of which was the Wildlife and Countryside Act of 1981, have attempted to redress the balance. Further protection followed from the Habitats Directive of 1992 of the European Union (EU). Legal protection from direct killing and other forms of exploitation currently covers a wide range of species considered at particular risk, and there are protected areas designated as Nature Reserves, Sites of Special Scientific Interest (SSSIs) and Special Areas of Conservation (SACs). While strengthening the law has certainly done some good, it has not proved to be anything like enough. Just as successive measures to reduce tax evasion have yet to convince critics of their effectiveness, so it is with wildlife legislation. Protection of species and habitats has all too often been flagrantly ignored or overridden by considerations alleged to be more important in the 'national interest', as if a healthy environment doesn't qualify for that assignation. A Secretary of State for the Environment in the 1980s was so incensed by his own department's successful challenge to a tax-avoiding forestry development in Scotland's Flow Country that he defenestrated an effective national organisation by splitting it into smaller, regional offices. Subsequent ministers have continued the trend of weakening the UK's statutory conservation bodies, hardly what we should expect from those with the power to make a positive difference. Raptor persecution on upland moors is illegal but regularly goes unpunished, and even the strongest habitat defences, such as SSSI designation, have been successfully challenged by developers. Perhaps the main failing, however, is simply the geographical extent of habitat protection on offer. More than 90 per cent of the UK countryside remains outside of SSSIs or any other defensive scheme for wildlife.

As a result of inadequate protection, wildlife declines have not been arrested, and the State of Nature reports have assessed the reasons behind this failure (Burns et al., 2016). Figure 2.5 summarises their conclusions, indicating not only the causes of decline but also their likely relative importance. Intensive agricultural practices, including the enormous increases in applications of fertilisers and pesticides since the Second World War, stand out as by far the most important change that has decimated the UK's wildlife. Hedges ripped out to create enormous

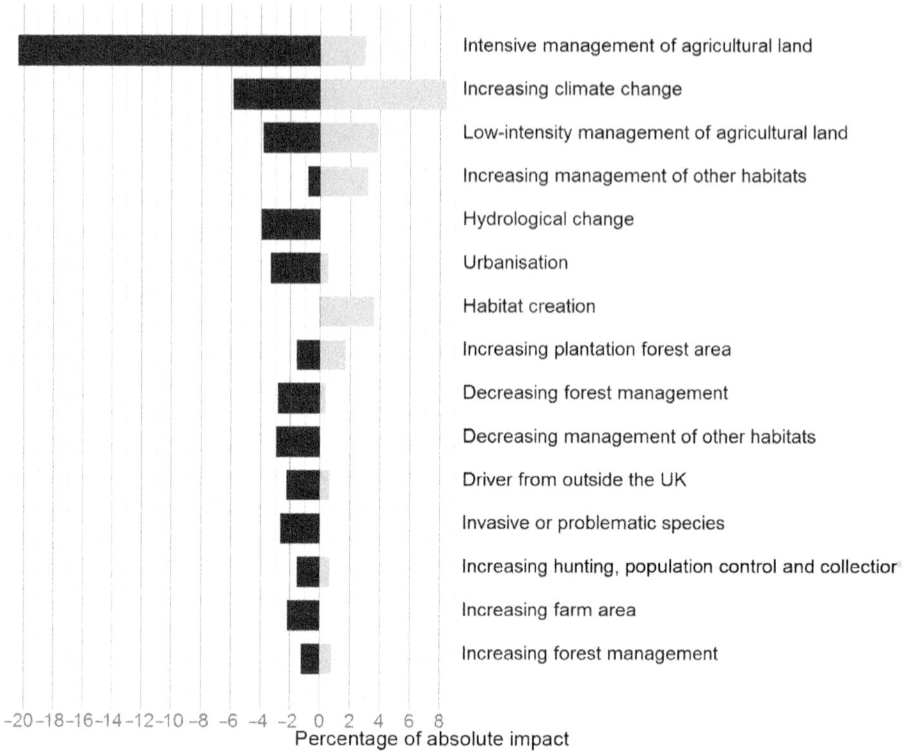

Figure 2.5 Major causes of recent wildlife declines. © 2016 Burns et al. under the terms of the Creative Commons Attribution License.

fields under arable cultivation, the sowing of rye grasses to 'improve' pastures, autumn rather than spring crop planting, use of silage rather than hay cuts, and neglect or removal of ponds have all contributed to the impoverished farmland wildlife we are left with today. On top of that, climate change has reduced the distribution or abundance of alpine plants and, especially, sea bird colonies that now suffer from depleted prey (especially sand eels [*Ammodytes tobianus*]) in warming seas. Freshwater habitats have been damaged by abstraction and pollution. Urbanisation spreads its tentacles over ever more green space, and some introduced species continue to wreak havoc with much-loved natives. Red squirrels (*Sciurus vulgaris*) are under continuous pressure from North American greys, and our native white-clawed crayfish (*Austropotamobius pallipes*) is in retreat as the signal crayfish (*Pacifastacus leniusculus*), also from

North America, ousts it from streams and rivers across Britain. Other contributory factors to the ongoing disaster include changes in management, for example, neglect of heathlands allowing scrub to overrun them, and increased hunting, persecution and collection of target species such as hen harriers (*Circus cyaneus*), almost resulting in their extinction in England. Not all recent changes have been deleterious; conservation-oriented habitat management, including countryside stewardship schemes, have had positive effects on wildlife, and even climate change has, thus far at least, not been entirely bad news for UK wildlife. Strong fliers able to cross the English Channel, mostly birds and insects such as dragonflies and damselflies, have increased biodiversity in our island archipelago. This trend is likely to continue, at least for a while, as the climate warms and becomes comfortable for previously 'southern' species. But most recent developments have been damaging to wildlife and there may be other, hidden factors not explicitly revealed by the State of Nature surveys. General disturbance, for example, is surely increasing in many habitats with usually unknown consequences for species exposed to it.

2.5 Not the Whole Story

All the reasons identified as problematic for wildlife in Britain have a common root, notably the activities, or occasionally the inactivities, of humans. Unlike some major events in prehistory, such as comet collisions or ice ages, no one is ascribing wildlife crises of the twenty-first century to agencies outside of human control. However, checking out the factors listed in Figure 2.5, we can look more deeply at what is behind them. In this book, I argue that most are merely symptoms, surely the immediate causes of wildlife declines but not the fundamental drivers of them. Consider a healthy river winding its way through a green and pleasant countryside. The wide variety of life in and around the water thrives on account of nutrients carried by it and which support a vibrant food chain from microorganisms through to fish, birds and mammals. However, this only works if the supply of nutrients remains within certain critical limits. If the situation changes to one of a nutrient overdose, the entire ecosystem can collapse. Such was the fate of many British rivers in the recent past, especially when untreated sewage was discharged into them from unregulated outlets. The River Trent was a prime example of this trauma. At one time an attractive feature of the North Midlands, in the seventeenth century renowned angler Izaak Walton considered it 'One of the finest rivers in the world and the most abounding with excellent salmon and all sorts of delicate fish.' By the mid-twentieth century, this situation had changed beyond recognition. At the end of the 1930s, a once bountiful run of salmon (*Salmo salar*) was no more. A sudden excess of nutrients provided by human waste led to long sections of river with few life forms other than sewage fungus and tubifex worms. Elsewhere there were mostly algae and water hog lice (*Asellus* spp.), while fish and higher plants were rare or absent altogether. All these symptoms of distress, changes in fauna and flora, and turbid, foul water resulted from one change, an overload of nutrients that in smaller quantities posed no problems to the water quality or the wildlife. Happily, the Trent is in a much improved condition today. Crude sewage (usually) no longer flows and salmon again leap over the Trent's weirs, a welcome indicator by this most sensitive of animals that in this case the worst excesses of nutrient overdose have been remedied.

The problems of rivers like the Trent illustrate a general point. With respect to the fate of British wildlife in general, we can translate nutrient overdose into excessive human numbers. A small human population is no problem and indeed benefits biodiversity by creating a range of different habitats. However, most of the activities damaging wildlife in Britain today are strongly influenced by a single primary cause, notably the sheer number of people now living on this small group of crowded islands. As with river nutrient overloads, the only sustainable, long-term solution for a healthy environment in the UK will require a reduction of the human population size. Subsequent chapters make the case for this claim, starting with what is perhaps the least important and leading up to the most damaging of all, intensive agriculture and massive levels of food production.

2.6 Overview

A large number of people enjoy and value wildlife in the UK. Natural history has a long and honourable past, and has been illuminated by a series of outstanding personalities over many centuries. Foremost among them was surely Charles Darwin, with a passion for nature that translated into an explanation of the natural world by one of science's most profound theories. Darwin was a naturalist first and foremost; one remarkable aspect of *The Origin of Species* is that it accomplished its goal with no mathematics whatsoever. Many naturalists today would relate to that, whereas scientists, including ecologists, rarely avoid an algorithm or two in any serious work. In a more economically obsessed age, it has also become necessary to put costs and benefits on wildlife. Although rigorous spreadsheets with plus and minus columns sufficient to satisfy accountants are currently beyond us, it is clear that our countryside has a substantive net value in both cultural and financial terms. Unfortunately, this has not proved sufficient to prevent ongoing, in many cases accelerating, declines of the animals and plants that decorate our British landscapes. The only hope of turning this situation around entails identification of not just the immediate causes of this problem, but also probing a fundamental issue that lies hovering in the mist behind them. That is what this book is about: people numbers.

3 · *Human Activities Directly Killing Wildlife*

3.1 The Physical Assault on Wildlife

In this chapter, the significance of human actions that directly cause the demise of wild plants and animals is considered. Here, as in other chapters, the extent and quality of evidence is crucial. We look to science for new ways of improving the health and wealth of humanity, and that ambition has very often been achieved. Advances in our understanding of the natural world are not, however, without price. For Francis Bacon, that Elizabethan architect of the scientific method, the cost of his innovation was eventually personal. He perished unsung from a chill acquired after collecting snow to stuff a chicken, all to test the effect of cold on the rate of decomposition. This far-sighted father of the enlightenment recognised not just the vast potential of scientific enquiry but also the dangers that future discoveries might bring. Today, we live in the shadow of these mixed blessings just as Bacon foresaw. Scientific research has propagated increasing arrays of specialised disciplines, sometimes generating contradictory outcomes. Conservation biology is a newish branch of the knowledge tree, one that we would be better off not needing, but increasingly essential because other offshoots, such as agrochemical developments, have grown strong enough to imperil the future of our countryside. Conservation-oriented research is the most robust way of investigating wildlife populations and of understanding the drivers of change. That virtually all wildlife declines in Britain are down to human activities of one sort or another is not contended. Of more than academic interest, however, are the mechanisms underpinning this sad situation. There are several of these, and in this chapter some of the most obvious ones are discussed, notably direct attacks on our flora and fauna mediated by collection, persecution and predation. As with all the issues described in later chapters, a critical aspect is the extent to which these assaults on animal and plant populations relate to the numbers of people involved.

3.2 A History of Attrition

Humans in Britain, as elsewhere, have a long history of wiping out wildlife. The first species thought to have disappeared from our islands as a direct result of human persecution was the auroch (*Bos primigenius*). These huge and formidable wild cattle trod pathways through a wild wood that spread across most of northern Europe as the Ice Age receded. They were probably lost from Britain well before the Romans arrived but persisted in the ancient woodlands of Eastern Europe until at least the seventeenth century. They then became globally extinct but have recently been resurrected by careful interbreeding of domestic strains of cattle, and auroch lookalikes now once again roam the Hungarian plains. Persecution in this instance meant predation; such large animals must have provided bountiful supplies of tasty steaks, albeit at considerable risk to hunters pursuing these large and aggressive beasts. A litany of extinctions in the UK followed over subsequent centuries including, in likely order of departure, brown bears (*Ursus arctos*), lynx (*Lynx lynx*), wild boar, beavers, wolves (*Canis lupus*), ospreys (*Pandion haliaetus*) and sea eagles. Bears, lynx and wolves (Figure 3.1) were dangerous nuisances, wild boar were good

Figure 3.1. Eurasian wolf, the last native mammal exterminated in the UK by human persecution. *Source*: Ginger/Moment/Getty Images (A black and white version of this figure will appear in some formats. For the colour version, please refer to the plate section.)

for eating, and beavers were prized for their pelts and scented castoreum secretions, while ospreys and sea eagles were predators of fish and lambs. For one species, persecution led to global extinction. The last great auk (*Pinguinus impennis*) in Britain was caught and killed on St Kilda in July 1844. That same month, the few remaining birds in the world, on Eldey Island near Iceland, were also dispatched. Three St Kilda men caught and restrained the final British auk, later killing it with a brick, believing it to be a witch that had caused a damaging storm.

The scale of wildlife persecution in Britain escalated enormously commensurate with the rising popularity of game shooting in the nineteenth century, itself largely consequent on the development of breech-loading shotguns. Pheasants (*Phasianus colchicus*), partridges (grey, *Perdix perdix*, and red-legged, *Alectoris rufa*) and red grouse (*Lagopus lagopus*) could then be murdered in huge numbers in a sporting activity that provided a major source of income for landowners of large estates. Just killing the game was not sufficient, and any predators inconsiderate enough to threaten the landed gentry's financial security had to go as well. Game keeping was the way to achieve that end, and by the end of the century, tens of thousands of men were gainfully employed trapping and killing foxes (*Vulpes vulpes*), stoats (*Mustela erminea*), weasels (*Mustela nivalis*), otters, badgers, various birds of prey and indeed any animal with carnivorous inclinations. We have the world wars to thank for relief from this form of attrition in the twentieth century, as social structures subsequently changed to decrease the affluence of upper-class hunting estates. One consequence of their downturn was a sharp decline in the number of gamekeepers. Even so, game shooting has remained popular and persecution of wildlife continues, often illegally, on many rural holdings across the UK. Happily, most victims of this Victorian killing extravaganza have recovered their numbers, although some species were driven almost to extinction by the start of the twentieth century. Polecats (*Mustela putorius*), pine martens (*Martes martes*) and Scottish wildcats (*Felis silvestris*) remained on the brink of disappearing from Britain for many decades and survived only in remote corners of Scotland, Wales or, more rarely, in the uplands of northern England. The recoveries of these animals have been slow, but at last more of us are enjoying the pleasure of their company as their ranges (apart, sadly, from the wildcat) grow strongly once again.

In past times, even naturalists were not averse to a bit of killing, especially when something unusual appeared on their patch. Gilbert White's passion for wildlife around Selborne was peaceful enough for the most part, although shooting crops up in his accounts when a close examination of some bird or mammal was deemed necessary. He recounts in *The Natural History and Antiquities of Selborne* that an osprey was shot (not by him) at Frensham Pond in 1772 'while sitting on the handle of a plough devouring a fish' (White, 1789). There was no thought for conservation in those days, and a common approach was to shoot first and ask questions later. In 1850, the Reverend Pickard-Cambridge's response to finding natterjack toads in a Dorset pond, and collecting a dozen of them, was not one of compassionate interest: 'I administered to each several smart blows with a walking stick, on which they quivered strongly. . .and became rigid, as though dead.' It turns out that they survived this assault, and also the subsequent administration of prussic acid, but succumbed to having their spinal cords cut with sharp scissors (Pickard-Cambridge, 1893–94).

The end of the nineteenth century witnessed the dawning of a new age in which wildlife persecution became, for the first time, worthy of disdain. Emily Williamson, disgusted by the enormous slaughter of birds to provide feathers for hats, founded the RSPB in 1889. Together with several other dedicated women, she inaugurated what was to become one of the most formidable and effective conservation organisations in the world. Even so, attitudes changed but slowly. Well into the twentieth century it was considered a suitable childhood occupation to raid bird nests and collect eggs, a hobby engaged in by several eminent naturalists who later became prominent conservationists. My youthful hunting instincts were directed mostly at pond life, the keeping of tadpoles, newts and aquatic insects in home-made aquaria. I would like to claim that this hobby was harmless, and certainly the intention was always to release the captives in due course. In effect, all too many of my collections resulted in animals eating each other, and for tadpoles in particular survival rates were probably lower than in the ponds from which they came. Nevertheless, encouraging children to catch and observe wild creatures in ways that do relatively little harm can only be a good thing. The Protection of Wild Birds Act in 1954, one of the first UK laws created to reduce wildlife destruction, made bird nesting illegal, and most of the parliamentary efforts to conserve wildlife since that time have focused on prevention of direct destruction or disturbance of particular species. This was commendable, of course, based on the history of persecution that had gone before, but it took some time to realise that in post-war Britain the situation had changed. Most species of animals and plants were not, and never had been, victims of intentional destruction. Imperilled habitats became the key issue for conservation, homes to huge numbers of different species and much more difficult to protect.

The question remains, though, as to what degree recent wildlife declines can be attributed to deliberate destruction despite the impressive battery of legislation now in place designed to protect many species. This problem has not entirely gone away.

3.3 Collection

Plants

Colourful fields in blooms and dashes of colour in woods and hedgerows are startling attractions of the British countryside. Temptation to pick some of the more eye-catching flowers can be overwhelming, and in most cases remains a harmless but nevertheless illegal activity unless sanctioned by the landowner. However, gathering wild plants at a sufficient scale as to cause local population declines has a long history in Britain and was particularly rampant in Victorian times. The lady's slipper orchid (*Cypripedium calceolus*) is one of the best-known victims of this collecting frenzy, and was declared extinct as a direct result of it in 1917. Happily, a single plant was rediscovered in Yorkshire in 1930 and subsequent captive breeding efforts have brought this attractive flower back from the brink. Another orchid, summer lady's tresses (*Spiranthes aestivalis*) (Figure 3.2), also fared badly and disappeared from Britain by the mid-1950s due to collection in the New Forest, its last remaining location in the UK. Other rarities have survived despite severe collection pressure, including the loss of more than half the known populations of lizard orchids (*Himantoglossum hircinum*). Fortunately, concern about plant conservation grew after the Second World War, and regulation of wild plant exploitation began in the mid-twentieth century. The taking of wildflowers was prohibited under the 1968 Theft Act, although it remains acceptable with owner permission. This kind of activity probably does little harm, but the removal of entire plants, rather than just flowers and leaves, is obviously more damaging and became illegal (again excepting with owner permission) under the 1981 Wildlife and Countryside Act.

The law has not put a stop to plant piracy. The conservation charity Plantlife (www.plantlife.org.uk) lists damage caused over the years by such inconsiderate and irresponsible behaviour, which is increasing despite the legal protection now in place. At the end of the twentieth

Figure 3.2. Summer lady's tresses, exterminated in Britain by specimen collection. Photograph by Hans Dekker, reproduced with permission from Saxifraga Foundation (A black and white version of this figure will appear in some formats. For the colour version, please refer to the plate section.)

century, a substantive series of plant thefts was continuing (Akeroyd, 1999). Whole-plant removal continues today and, although less important than habitat destruction, it has certainly contributed to recent declines of some species. The dark red helleborine (*Epipactis atrorubens*) is a rare inhabitant of limestone hills and pavements in northern Britain. In 2008, thieves removed their glorious displays from the Peak District National Park. Ongoing collection still has the potential to cause extinction. The only known specimen of a hybrid spleenwort fern (*Asplenium* spp.) was stolen from a wall in Staffordshire in 2000. Not only are rarities at risk. Large numbers of snowdrops (*Galanthus nivalis*), common bluebells (*Hyacinthoides non-scripta*) and wild garlic (*Allium ursinum*) plants have been stolen for commercial profit, often ending up for sale on eBay. The UK is home to almost half the global population of common bluebells, but this signature of spring is declining due to loss of woodland and, especially, to hybridisation with the non-native Spanish bluebell (*Hyacinthoides hispanica*). Extra pressure from illicit, large-scale collection is not what this azure beauty needs. At just one site in Cumbria, 5,000 bulbs were taken over 2 years. Evidently, plant theft can be big business, and police have identified gangs operating in East Anglia using specially adapted tools to dig up their target species. This is a crime not easily detected, and very likely much more of it goes on than ever makes it to the courts. The average of five incidents of plant theft detected annually in Cambridgeshire alone in the 1990s is surely just the tip of an iceberg.

There are two dimensions to the problem of wild plant collection. How many perpetrators are out there and, perhaps more significantly, how large is the market for their illicit wares? In neither instance are robust statistics available. Court cases and police reports do no more than indicate minimum numbers of thieves, while there is next to no information about those receiving the stolen goods. Evidently, plant collection has transformed from a primarily private obsession in the nineteenth century into what has become very largely a commercial business. Even without hard data, it seems likely that the numbers of people involved have increased at both the supply and demand ends of the enterprise. Arguably, at least, the crime has probably magnified according to the increasing human population in place to nurture it.

Invertebrates

Like most of Britain's wildlife, many species of invertebrates are in decline, and insects in particular are experiencing catastrophic losses. Most of the damage has been done by agricultural intensification (see Chapter 5), but in a few cases, collection and predation have had notable impacts. The large blue butterfly died out in Britain in the 1970s, but, after ground-breaking research to identify the reasons for its demise, it was subsequently reintroduced to England with marvellous success. Nevertheless, it remains a rare and localised beauty confined to a few select, well-managed locations, making it a target for selfish collectors with attitudes reminiscent of Victorian hobbyists. In 2017, Phillip Cullen was convicted of catching and killing not only large blues but several other rare butterflies to stock a private collection. A huge amount of effort has gone into the large blue success story, but while the insect remains rare, which it probably always will, specimen collection could have a negative impact on small populations. As we know from ongoing prosecutions of bird egg collectors, cavalier nineteenth-century attitudes to collecting rarities have not completely died out. Entomologists continue to make collections of pinned specimens, quite legally with regard to most species, for personal identification guides. There is no evidence to suggest that most collection of invertebrates is problematic for conservation, perhaps excepting a few very rare species.

The Ongoing Impacts of Collection

The taking of wildlife for profit or personal collection therefore continues, in some cases at a scale similar to or greater than that of Victorian times. Fortunately, though, the ethos has largely changed, partly because of protective legislation but also due to an increasing awareness of the need for conservation. However, there are still people in the business of taking rare plants and invertebrates as well as the eggs of wild birds, and it remains necessary to monitor and protect the most sensitive places where these species persist. It was considered essential as recently as 2010 to defend one of the last surviving lady's slipper orchids by 24-hour police patrols. This single plant had been mutilated at least twice in the previous 6 years by collectors intent on taking cuttings. Such selfish individuals need to be deterred because it is mostly species with tiny populations that are at risk of decline or extinction due to irresponsible harvesting. A few plants, especially orchids, remain imperilled by these uncouth activities as, just possibly, do butterflies like the large blue.

More generally, the overall level of international wildlife trade continues unabated and indeed quadrupled between 1975 and 2014 with, by that time, the equivalent of 100 million individual organisms per year changing hands (Harfoot et al., 2018). Wildlife crime has continued to increase in the UK, by 9 per cent between 2017 and 2018 (www.wcl.org.uk/wildlife-crime.asp). Orchid collection remains rampant, with millions of plants traded across the world every year. While most specimens are artificially propagated, up to 10 per cent are stolen from the wild. After a decline in the proportion of wild-caught plant sales at the start of the millennium, numbers have risen continuously since 2008 (Hinsley et al., 2018). These include unknown numbers of illegally collected rarities, providing for an increasing international market. Butterfly collection also remains problematic because substantial sums of money can change hands. The conviction of Phillip Cullen for taking specimens of the large blue was an unusual instance of someone being apprehended for a crime that is all too easy to get away with. This species can sell for up to £300 when mounted to simulate legally taken Victorian specimens. In 2017, swallowtail butterfly (*Papilio machaon*) caterpillars were stolen from Hickling Broad nature reserve in Norfolk. This is another rare species with a very restricted distribution on the Norfolk Broads. Collection for the wildlife trade is, on a global scale, considered by the World Wide Fund for Nature (WWF, formerly the World Wildlife Fund) to be an important cause of species declines, second only to habitat destruction (www.worldwildlife.org/threats/illegal-wildlife-trade): 'As human populations have grown, so has the demand for wildlife. People in many countries are accustomed to a lifestyle which fuels demand for wildlife. They expect access to a variety of sea foods, leather goods, timbers, medicinal ingredients and textiles. At the other end, extreme poverty means some people see wildlife as valuable barter for trade.' Hard data about the extent of wildlife commerce are available for very few countries, but the USA is an exception (Smith et al., 2017), and Figure 3.3 shows how trade and human population size both increased over the first decade of the twenty-first century. However, trade, measured in hundreds of thousands of individual plants and animals, rose faster than the human population consuming it. The ratio of imports to people increased more than 1.5-fold over this period, indicating that higher consumption rather than population size was the main driver of wildlife exploitation in this case. The international illegal wildlife trade is reckoned to be worth some £19 billion per year, beaten in value only by those in drugs, counterfeiting and human trafficking. This debilitating situation is proving very hard to counter.

3.4 Persecution

Pest Control

Deliberate destruction of wildlife remains rampant in the UK, mostly targeted at species considered to be problematic nuisances. Gardens provide startling examples of pest control. Lovers of the perfect lawn as an unbroken green sward wage war against daisies (*Bellis perennis*), dandelions (*Taraxacum officinale*), buttercups (*Ranunculus* spp.), mosses or any other plant that attempts to make a dull grass carpet more interesting. Flower beds attract even more attention with the intrusions of 'weeds', wild insurgents that deny the gardener neat and tidy borders. Attrition

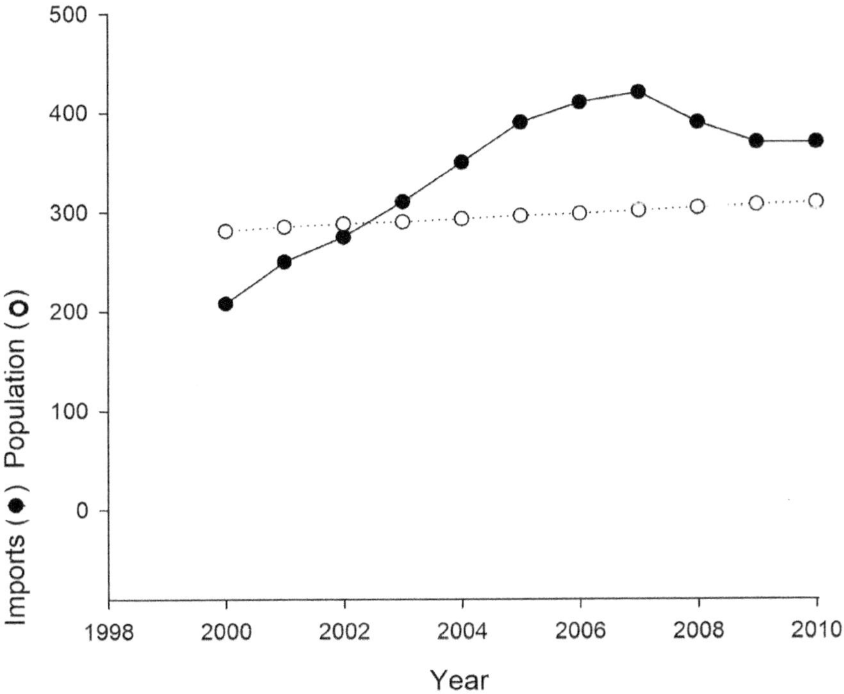

Figure 3.3. Trends in wildlife trade and population size in the USA. Imports are in thousands of plants and animals; human population is in millions.

may be physical or chemical, using the wide range of tools and herbicides on sale at garden centres. Then there are unpopular invertebrates, aphids feeding on rose bushes, and snails and slugs rampaging through the succulents. Often chemicals again come to hand, and some are potent toxins with highly undesirable consequences for non-target species. Slug pellets containing metaldehyde have damaging effects further up the food chain when poisoned molluscs are consumed by small mammals (especially hedgehogs) and birds such as song thrushes (*Turdus philomelos*). Mortality of these unintended victims can result from multiple organ failure. Fortunately, metaldehyde will be banned from garden use in 2022. Chemical weapons are commonplace within houses as well as outside them to combat outbreaks of woodworm or, more often, to spray flies and wasps. For many of the annoyances faced by gardeners, alternatives to chemical defences are available, including physical protection of sensitive plants by materials that slugs and snails are reluctant to cross. With the importance of gardens as wildlife refuges increasingly recognised, more efforts should surely be made to move away from chemical warfare on our home grounds.

The fight against undesirables extends way beyond the garden into the open countryside. There is a perpetual war against rodents, based mostly on physical trapping and poisoning, with mice and rats as the primary targets. As with slug pellets, some rodenticides can have damaging effects on predators such as barn owls (*Tyto alba*), which find sick animals easy victims. This is a war that can never be won, so unrelenting efforts of containment are the order of the day. It is estimated that Britain is home to many millions of brown rats, and their numbers continue to increase in defiance of all attempts at control. Rabbits and brown hares (*Lepus europaeus*), too, are widely persecuted despite substantial ongoing declines of both species. Plenty of birds also have a rough time, particularly those minimally protected by law and which many farmers look on as costly enemies. Pigeons and corvids are particularly disliked, and many are shot on a regular basis to protect crops.

Those plants and animals unluckily designated as pests continue to recoil at the hand of man, but most of the species involved attract little sympathy and are unlikely to suffer population declines consequent upon the attrition foisted on them. Nevertheless, some, including rabbits and hares, are loved by very many people. They are in trouble mainly for reasons unrelated to direct persecution, notably habitat changes and rampant diseases, but persecution is an added pressure that can only make their situations worse. In these instances, one person's pest is another's delight, and definitions become seriously contentious.

Problems or Scapegoats?

Wildlife persecution is not limited to species that many people are happy to see controlled. Badgers are perhaps the best-known example of a charismatic, much-loved animal recently subjected to mass killing in an attempt to reduce outbreaks of bovine tuberculosis (bTB) in cattle herds. Thousands of badgers have been shot in selected areas of (mostly) south-west England, and these control zones have been progressively expanded. Thus far, the evidence suggests that badger culls have had at best a small effect on the incidence of bTB, and unreliable tests for the causative bacterium, poor farm hygiene and cattle movements are much more important as perpetuators of the disease. Killing wildlife is nevertheless the option favoured by many farmers, apparently unprepared to tackle the other, maybe more expensive, disease control options. An awkward dimension to this dilemma is the relationship between badgers and hedgehogs. The evidence is contentious, but research has linked hedgehog declines at least in part to an increasing badger population. Badgers definitely eat hedgehogs sometimes, and I have seen an inside-out hedgehog skin in a Sussex garden, which is the typical outcome of badger predation. However, badgers have a very varied diet, and evidence suggests that hedgehogs are usually rare items on their menu. In gardens, both species have been filmed feeding peacefully together and of course they have coexisted in the UK for many thousands of years. It may be that intensive farming has removed so much good habitat that badgers and hedgehogs compete more intensively than was previously the case. In our garden, we never see hedgehogs now, but badgers visit regularly, and there is anecdotal evidence that hedgehog numbers have recovered somewhat in badger cull zones. The jury is out on this one, but it highlights the complexity of ecological interactions between two species both of which most people like to have around. Fortunately, badger numbers are buoyant, and population declines resulting from the culls are unlikely to threaten the species in the UK. Culling has been a cruel and unpalatable example of political expediency overriding scientific advice, but at the time of writing there is better news. Badger killing as the main approach to contain bTB may be phased out in England over the next decade and replaced with a long-overdue vaccination programme.

A paradoxical and contentious form of persecution is occasionally carried out by conservation organisations. Efforts to arrest the long-term decline of Britain's red squirrels have frequently invoked trapping and shooting grey squirrels, those North American invaders that have progressively ousted our native species over most of England and Wales. This is a perpetual task that has achieved some local successes, but could by itself never eliminate the alien from our shores. A more interesting solution to the problem stems from the discovery that pine martens predate grey squirrels more effectively than they do reds. This graceful mustelid is gradually recovering its historic range, sometimes with help from human enthusiasts since legal protection reduced its persecution by gamekeepers. At least in areas of woodland large enough to support it, the return of this partially selective predator offers some long-term hope for red squirrels. Efforts have recently begun to bring pine martens back to the Forest of Dean, where they will join a few beavers, a burgeoning population of wild boar and maybe, one day, returning red squirrels. With luck, this natural control method should relieve the bad press that killing the damaging but delightful grey invaders by conservationists inevitably attracts. Other aliens, generally less popular than squirrels, have also been subjected to attempts at eradication. North American mink (*Neovison vison*), escapees from fur farms, have spread across most of Britain and left a trail of devastation in their wake. A competent swimmer, this efficient killer has raided the nests of ground-nesting birds on islands previously safe from most predators. Even more seriously, it is mainly thanks to mink that water voles have changed their status from one of abundance almost everywhere in ponds, ditches and canals to become the UK's fastest declining vertebrate. In the 1970s, water voles were in almost every waterway where I went pond dipping, from abandoned canals to heathland bogs. Now, hearing the distinctive plop as one dives beneath the water is a rare and precious experience. Mink are now trapped and killed mercilessly throughout the land, leading to some local recoveries of water vole populations. The good news is that, as with red squirrels, a native predator is coming to the rescue. The recovery of otter populations from pesticide poisoning has, in recent decades, seen them regain most of their old haunts, and their reappearance seems linked, at least sometimes, to mink declines. There is talk of what until recently seemed an impossible task, the complete eradication of mink in Britain. It is a tall order, but we shall see. There are precedents. North American muskrats (*Ondatra zibethicus*) and South American coypu (*Myocastor coypus*) are large rodents that did very well after escape from farms in the UK, but both were completely extirpated by intensive trapping before the end of the twentieth century.

Most people can probably accept the persecution of alien species that threaten natives or, in the case of the two American rodents, damage river banks and reed beds. However, the destructive activities of conservationists have not stopped there. Predation of ground-nesting birds has prompted highly respected conservation bodies, including the RSPB, to sanction the killing of animals on nature reserves and offshore islands. Fox culls have occurred or are planned at nature reserves in Wales, northern England, Scotland and Northern Ireland. The main aim of this exercise is to increase survival of curlew chicks, a species endangered primarily by changed farming methods and which needs desperate measures to prevent further declines. In all cases, culling is a measure of last resort, and methods of deterrence, such as electric fences, are employed before anything more drastic is done. Foxes are common across most of the UK, and populations are unlikely to suffer much damage from highly localised culls. Even so, some people are dismayed that conservation bodies, however well intentioned, are shooting animals rather than protecting them. The RSPB is transparent in its reporting of this work and unrepentant about the need for it, but it is certainly not ideal for public relations. Even more contentious was the mass lethal injection of hedgehogs on the Outer Hebrides, again as part of a plan to reduce predation of ground-nesting birds. Hedgehogs are not native to these Scottish islands, having been introduced by humans, and threatened the survival of increasingly rare birds including dunlins (*Calidris alpina*), snipes (*Gallinago gallinago*), ringed plovers (*Charadrius hiaticula*) and redshanks (*Tringa totanus*). The hedgehog cull, originally sanctioned by Scottish Natural Heritage, was terminated following much adverse publicity and replaced by a policy of live capture followed by release on the Scottish mainland. In practice, this strategy probably also amounted to a death sentence, as the translocated animals must have struggled to survive in habitats already well populated with indigenous hedgehogs. Subsequent deaths from starvation would have been out of sight and out of mind. Hedgehog control has proved even more controversial than that of foxes because the spiny mammals are hugely popular and because they have declined appallingly in recent decades. These issues pose serious dilemmas for conservationists trying to preserve what is left of our wildlife and simultaneously increasing public support for the task. There is no doubt that predator control has been well intentioned, undertaken only when all else fails, and has usually proved effective in attaining its objectives. In no cases, even with hedgehogs, have the local control initiatives contributed significantly to national population reductions. Nevertheless, culling remains an unhappy backstop for wildlife conservation and is necessary mostly because primary causes of declines, especially habitat destruction, have left some species requiring special measures if there is any hope of maintaining their presence in Britain.

Business as Usual

Then we come again to shooting estates and the role of the gamekeeper in twenty-first-century Britain. Although far fewer than they were a century ago, these defenders of partridges and pheasants continue their attrition against weasels, stoats, foxes and other potential predators across much of lowland England. Astonishingly, licences have been granted to shoot buzzards, a major recent conservation success story, where these are considered a threat to landowners' profits. Fortunately, these activities have not impacted significantly on populations of the target species, most of which survived the best efforts of gamekeepers when there were many more of them about. Of greater concern is the damage done by the release of huge numbers of game birds, mostly pheasants, into the countryside. More than 35 million are released annually, with potentially serious consequences for native species including lizards and snakes upon which they prey. It seems outrageous that release of a non-native bird on this enormous scale is permitted at all, somehow escaping legislation that prohibits introduction of alien species into Britain. That this behaviour has been supported by legal destruction of a recently recovered raptor beggars belief. Moving uphill, the persecution of birds of prey and mountain hares (*Lepus timidus*) on grouse moors is even more shocking. Illegal poisoning, trapping and shooting on these estates regularly make the headlines but have, thus far, failed to curtail the activities of the perpetrators where detection of their crimes is difficult to prove. The RSPB and other conservation organisations do their best to address this problem and there have been successful prosecutions of rogue gamekeepers, but fines are usually desultory and the carnage continues. It is likely that golden eagles (*Aquila chrysaetos*) would have spread beyond northern Scotland if the invisible barrier of illegal persecution on grouse moors did not stand in their way. In England's Peak District, eagle owls (*Bubo bubo*) and goshawks (*Accipiter gentilis*) have been victims of over-zealous keepers, but particularly dismal has been the annihilation of hen harriers in England (Figure 3.4). Attempts to breed by this haunting raptor on northern moors have been thwarted year on year, sometimes so successfully that no young survive. And damage goes beyond the overt killing. Moors are deliberately burnt to encourage the growth of new heather upon which grouse feed. This practice depletes the underlying layer of peat, a major carbon store, which is invaluable as a buffer against climate change. It also accelerates run-off of water after heavy rainfall, a common event on upland moors, with consequent increased flood risk at lower altitudes. The rate of burning has doubled over the past 50 years, and moorlands have become carbon producers rather than the carbon sinks they maintained in the past. Some good news is that burning was banned in several sensitive areas in 2020 and a general prohibition is on the cards. Estate owners can hardly be surprised that moves to license driven grouse shooting are gaining traction. This is one area where persecution is clearly still affecting some much-loved members of the UK's wildlife heritage.

Figure 3.4. Hen harrier killed on a shooting estate. With permission from RSPB

Extent of Persecution

Game shooting is on the increase in Britain, with ever more people taking up this 'sport'. Thousands of children under the age of 17 held shotgun licences in 2016, and numbers were increasing by more than 5 per cent annually. It is estimated that by 2013 at least 600,000 people in the UK were taking aim at live quarry, clay pigeons or targets, and at least 1.6 million individuals were shooting animals with airguns (www.shootingfacts.co.uk). Advocates of these activities emphasise the positive aspects of the hobby, such as the amount of money spent on supporting goods and services, jobs in the industry and conservation benefits from land management favouring game survival and, in some but not all cases, native wildlife. However, the sheer numbers of people involved are used to justify the massive numbers of pheasants released annually with destructive consequences for native species. Similar pressures apply on upland estates to provide grouse whatever the cost to rare raptors. According to the British Association for Shooting and Conservation, around 40,000 people visit grouse shooting estates annually, and pay large sums for the privilege of blasting away at the birds (British Association for Shooting & Conservation, 2015). Securing this income requires sufficient numbers of game to satisfy these demanding customers, and woe betide any predator that interferes with the cash flow. A particularly shocking aspect of large-scale pheasant shooting is the wastage involved. Many carcasses are simply buried in the field rather than being sold on for food. Again it is a question of scale; more people mean greater pressures on vulnerable wildlife and increased rates of species declines. Arguably, the most serious example of extinction risk from this pursuit today is the hen harrier in England. If the current level of abuse by moorland estates continues, the country could lose its few remaining pairs.

3.5 Predation

Invertebrates

Humans have predated wildlife in Britain from time immemorial, probably beginning with the arrival of archaic peoples nearly a million years ago and continuing intermittently ever since. For most of the intervening period, human numbers were too low to have serious impacts on wildlife populations, but in post-glacial times things began to change. As mentioned earlier, aurochs were perhaps the first animals exterminated by humans in Britain more than 2,000 years ago. In the twenty-first century, life is very different from that distant past, and people in the UK no longer rely on wild animals for food. However, predation is not just about nutrition. A dramatic example of invertebrate destruction for minerals rather than for meat is that of freshwater pearl mussels (*Margaritifera margaritifera*). Serried ranks of these molluscs bedeck the beds of upland Britain's most pristine rivers. Their life history entails the production of tiny, free-swimming larvae termed glochidia, which, if they are to survive, must latch onto the gills of passing salmonid fish. Here, they benefit from physical protection as they grow, eventually bailing out to begin a sedentary life on the rocks and gravels of the river floor. This chancy lifestyle means that survival to adulthood is low, to some extent balanced by a long lifespan (up to 100 years or so) for those that do make it. But pearl mussels are in serious decline. In part, this is because many rivers they once frequented are no longer in good condition, suffering from pollution or, especially, from increasing siltation, which creates unsuitable conditions on the river bed. Because mussels are so long-lived, it took some time to realise that many of the remaining populations were not recruiting new members but were constituted solely by ageing adults. At least pollution can be remedied and water quality in some rivers has improved, but this has not been the mussels' only problem. They share an interesting but damning feature with marine oysters, notably the ability to cope with irritant particles swept into their shell by secreting a calcium-rich covering to envelop the interloper. In this way, pearls are formed by what is essentially a molluscan immune response. Most individuals never produce them, but this has not prevented mass exploitation by illegal mussel fishing, a process that involves harvesting whole populations, opening their shells and leaving the resultant carnage in piles on the river bank. Because mussel beds are readily accessible, these massacres are all too easy to perform, and no matter that the vast majority of victims produce nothing of value. Such wanton exploitation has been a major cause of pearl mussel declines, and because remote areas are often the sites of this crime, it has proved singularly difficult to contain the destruction. Mussel pillaging accelerated towards the end of the twentieth century, and the animals were exterminated in an average of two Scottish rivers per year between 1970 and 1998. Legal protection was granted in 1998, but illicit persecution continues. Thousands of discarded shells were recently found in a single river in Assynt, north-west Scotland, and in 2018, 80 pearl mussels were killed by poachers on the river Spey. Pearl mussels are internationally endangered and are probably surviving in only 50 or so rivers outside the UK, mostly in northern Europe and Canada. And so it goes on. Human predation is having a serious impact on one of the UK's rarest and most interesting invertebrates. From 2013 to 2016, there were still, on average, two recorded incidents of pearl mussel crimes per year. No doubt many more went unnoticed.

Another invertebrate subject to large-scale human predation, in this case legally, is the common lobster (*Homarus gammarus*). These armour-plated crustaceans lurk in crevices, and among rocks, beneath pounding seas all around the UK. Much prized in restaurants everywhere, the common lobster is fished commercially pretty much wherever boats take to the water. Fresh animals were selling for £20 or more each in Cornish markets in 2018, making lobster catching a highly lucrative business. Despite the high cost of a tasty meal, demand for lobsters has soared and 170 tonnes, worth £2.8 million, were landed in Britain in 2017. Unfortunately, the popularity of these alien-looking submariners has had consequences, and there is increasing concern that stocks around British coasts are declining. This is despite guidelines that require lobsters below a minimum size to be released immediately after catch, and in response to stock concerns, DEFRA has consulted on rules to return 'berried' animals (egg-bearing females) as well. Current levels of exploitation are unlikely to be sustainable, and periodic sharp declines have been noticed in Britain and Scandinavia. Norwegian catches fell from around 1,000 tonnes annually in the 1960s to as few as 30 tonnes in the 2010s. Apart from the fishing guidelines mentioned above, additional measures are also now in place to improve lobster stocks. Artificial breeding centres, including the National Lobster Hatchery in Cornwall, release tens of thousands of captive-reared young lobsters into the sea every year. Marine nature reserves act as safe houses for many species, including lobsters, by enforcing no-take fishing zones. As demonstrated by the reserve around Lundy Island, this recent conservation measure can have rapid beneficial effects, and lobsters around the island are already larger and more abundant than before fishing was controlled. Unfortunately, such marine sanctuaries are few and far between, and demand for lobsters increases despite the high prices the animals command.

Vertebrates

In terms of sheer quantity, human predation of wildlife in Britain today is most pronounced in the case of marine fishes. Consumption of these animals continues to rise year on year, accentuated by the growing human population, and is expected to be more than 9,000 tonnes per week in 2030 (https://thefishsite.com). Fish that have proved the most popular over time are Atlantic cod (*Gadus morhua*), haddock (*Melanogrammus aeglefinus*), plaice (*Pleuronectes platessa*), Atlantic mackerel (*Scomber scombrus*), trout (mostly rainbow (*Oncorhynchus mykiss*) but also *Salmo* spp.), Atlantic herring (*Clupea harengus*) and pollock, including coley (*Pollachius* spp.). How are these largely invisible members of UK wildlife faring under this intense pressure?

The Marine Conservation Society (www.mcsuk.org) warns that despite regulatory efforts the Atlantic cod remains in crisis. Most cod stocks in the north-east Atlantic including those in British waters are overfished, inefficiently managed or at an unknown level. Those in the Irish Sea have recently fallen drastically. The European Environment Agency reckons that the risk of a cod population collapse in the North Sea remains high, and that the stock remains below safe biological limits. Sole around Britain's west and south coasts have been overfished almost to the point of collapse, according to the UK's Biodiversity Action Plan. On a happier note, plaice stocks appear to be holding up after severe declines noted in the 1970s and 1980s, as do those of mackerel, herring and coley. Catch limits and other conservation measures implemented since 2000 have had significant beneficial effects and have stabilised at least some marine populations (e.g. ICES, 2020). Nevertheless, the total amount of marine fish landed in Britain has declined consistently over the last 40 years, overall by about 60 per cent (Uberoi et al., 2021). The main reduction since 1998 has been of species that live on or near the sea floor, especially whiting, haddock and cod. Those swimming higher in the water column, such as mackerel, have proved more resilient although sporadic declines of these species have also occurred. Both the size of the UK fishing fleet and the numbers of people employed in the fishing industry have declined commensurate with the decreased catches. Before 1983, the UK was a net exporter of fish, but since that time imports have played an increasing role in supplying customer demand. In 2016, UK vessels landed some 450,000 tonnes of fish in Britain while in the same year more than 700,000 tonnes were imported. Fish stocks, at least of some key species popular with consumers, declined severely towards the end of the twentieth century (Figure 3.5) and British people now rely increasingly on imports. Taken together, these trends imply that predation has played a significant role in downward trajectories of several marine fishes native to British seas. There are precedents for fishery disasters in British waters. The magnificent common sturgeon (*Acipenser*

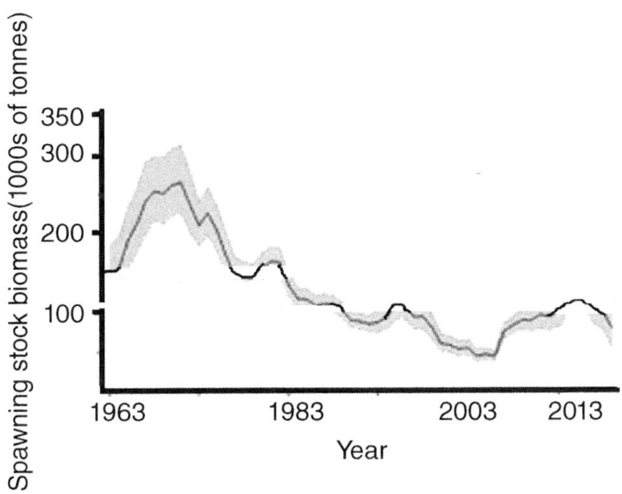

Figure 3.5. Changing catches of cod over 50 years. Based on data from ICES (2020)

sturio) was once a regular visitor to UK rivers and was sufficiently common in the North Sea to support a substantial fishing fleet. Now these huge animals are rarely seen, let alone caught, around our shores, depleted beyond redemption by unregulated catches that ceased in the early twentieth century when there were too few left to warrant the effort. Sturgeon caught in UK rivers are the property of the reigning monarch and must be offered up accordingly, but it is highly unlikely that the current or future incumbents of the British throne will have opportunities to appreciate such a gift.

Although of lesser concern than what is going on beneath the waves, terrestrial vertebrates do not escape unscathed from the predatory instincts of *Homo sapiens*. However, in these cases, conservation concerns mostly involve inadequate control rather than overkill. Deer numbers in the British countryside are thought to be at their highest level since the Norman invasion, and have doubled during the twenty-first century to stand, it is estimated, at around 2 million of the six species – roe (*Capreolus capreolus*), red (*Cervus elaphus*), fallow (*Dama dama*), sika (*Cervus nippon*), Reeves's muntjac (*Muntiacus reevesi*) and Chinese water (*Hydropotes inermis*) – combined. At present, some 350,000 animals are culled annually. Sales of venison in the UK have risen very rapidly, although much of this comes from deer farms. Despite a high and increasing level of predation, there is no reason to suppose that harvesting wild deer is impacting adversely on their populations. Indeed, a case can be made for increasing the cull, especially as only two species (red and roe) are native British animals. In some places, the surge in deer numbers has caused serious damage to trees and to understory plants in forests including precious bluebell stands. They are also a major reservoir of Lyme disease-bearing ticks, an illness that is on the rise in the UK. Game butchers regularly offer a range of other species including rabbits, wood pigeons (*Columba palumbus*), wild mallard (*Anas platyrhynchos*) and various other birds including pheasants and partridges, but the popularity of these strongly flavoured meats remains relatively low among British diners, and none is threatened by human predation.

Impacts of Predation

Excessive predation sufficient to endanger wildlife relates mainly to rarities like pearl mussels and more widely to marine species around Britain's coast that are taken for food, a vast harvest that continues largely unseen beneath the waves. Overfishing impacts not only on those species of fish and invertebrates prized for the table but also incidentally on seabed biota in habitats wrecked by intensive fishing methods, as described in Chapter 5, and further up the food chain it damages sea bird populations. This large-scale plundering of the seas around our shores is certainly the most substantive aspect of the direct killing that humans impose on British wildlife today. Despite evidence of stock depletion and efforts to regulate fishing, impacts on fish populations continue to escalate and are clearly influenced by the increasing human population. According to the Food and Agriculture Organization of the United Nations (FAO, 2016), the percentage of fishery stocks harvested at biologically unsustainable levels increased from 10 per cent in 1974 to 26 per cent in 1989. After 1990, this trend continued, albeit more slowly, and, conversely, fish stocks within sustainable levels dropped from 90 per cent of the global total in 1974 to 69 per cent by 2013. Thus, by the start of the twenty-first century, 31 per cent of fish stocks were overfished. Most stocks were 'fully' fished by 2013, meaning there was no potential for increases in production; the remaining stocks were already overfished and set to decline. During this same period, per-capita fish consumption increased steadily around the world, partly because the health benefits of this protein source have become widely recognised. However, unlike the situation with wildlife imports (Figure 3.3), the increases in marine fish harvesting relate more strongly to human population size than to the relatively small changes in individual eating habits.

Evidently two related factors threaten the future of wild marine fish around Britain's coast. Both increasing human populations and, to a lesser extent, rising individual consumption correlate with wild global fishery declines. Loss of fish species in Austrian lakes has been linked directly to human population increases (Zick et al., 2006). Similarly, declines and extinctions of large fish in the Caribbean correlate with increasing human population densities (Stallings, 2009). Thus, increasing human population and demand have been widespread primary causes of fish declines and this situation looks certain to persist. Ever-increasing demand has also generated inadequately regulated aquaculture (see Chapter 5), which can make the conservation situation worse rather than better.

3.6 Overview

Hundreds of thousands of plants and animals are destroyed deliberately by humans every year in UK houses, gardens, countryside and neighbouring seas. To what extent do these losses, brought about by collection, persecution and predation, contribute to overall wildlife declines in the UK? And to what degree is the attrition related to the size of the human population? The scale of killing is in some cases high, and collateral damage such as destruction of sea bed biota by aggressive fishing techniques accounts for many more losses than are readily documented. Many victims of targeted destruction are abundant species widely regarded as pests, for which few tears will be shed. Slugs, snails, rats, houseflies and the like fall before us in huge numbers with no detectable impacts on their populations. Species perceived as problematic or delightful, depending on your point of view, include grey squirrels, deer, badgers and foxes. On the basis of current evidence, wars waged against these stalwarts of our countryside are unlikely to threaten extinction, or in most cases even significant declines. It seems fair to say that although the direct killing of wildlife by humans continues apace it is, on balance, of relatively low impact compared with other damaging activities in the countryside described in subsequent chapters. There are, however, two groups for which persecution is certainly not inconsequential. These are rarities for which even a small number of deaths can have an adverse effect on their populations and popular prey species subject to heavy, probably unsustainable, harvesting. Raptors and, especially, out-of-sight marine species have taken big hits from human attacks, and an increasing human population in the UK has ensured that the killing continues ever more effectively. Fewer people on the grouse moors and less demand for fish are the best, arguably the only, way to reduce this pressure. But fish are a nutritious, healthy food and it is hard to see how reducing demand can be achieved without reducing human numbers as well.

Evidence for the role of the human population size on the various types of direct killing described in this chapter is variable. If the USA is typical, wildlife collection and sale has increased mostly as a function of higher per-capita consumption, presumably reflecting increased affluence in the receptor countries, with a relatively minor component attributable to population growth. But there is a much stronger relationship between marine fish stock depletion and human numbers. The fish depredations have been large and constitute a significant component of British wildlife declines.

4 · *Impacts of Development on Wildlife Declines*

4.1 Beware the Urban Jungle

Perhaps the commonest perception of threats to the British countryside is that of building over too much of it. This is scarcely surprising. The pressure for increases in housing seems relentless and has been ongoing for decades, with ever more land submerged beneath concrete and tarmac. The experience of watching green fields disappear forever is commonplace cross the UK, and where I live in Somerset, there seems to be a new housing estate raising its roofs against the skyline every year. Altogether about 18,000 km^2 of the British landscape, approximately 8 per cent of the total, was classified as urban in 2017 (www.ons.gov.uk/economy/environmentalaccounts/bulletins/uknaturalcapital/urbanaccounts). About 85 per cent of this is in England, meaning that in this member of the kingdom as much as 16 per cent of the country is no longer rural. Between 1990 and 2015 alone, urban land cover increased by 3,376 km^2 (www.ceh.ac.uk/ukceh-land-cover-maps), which is approximately 1.4 per cent of the total UK land area. Again development has been disproportionate, with about twice as much in the places where most population growth is occurring, notably lowland England, south Wales and the central belt of Scotland, compared with the less populous highland regions. Therefore, almost 19 per cent of the urban sprawl present in 2017 has been generated since 1990, and that is a lot of countryside to lose in a mere quarter of a century. Comparisons with other west European countries are not straightforward because each has its own definition of urban space, but England is clearly at the top of the league table together with Belgium and the Netherlands. Sadly, people opposed to this trend, especially when development is close to home, are commonly derided as NIMBYS ('not in my back yard') rather than protectors of green space. In recognising this conflict of interests, an attempt to alleviate the pressure of building on the countryside was initiated with the concept of greenbelt, a ring of land around towns and cities in which development is restricted or prohibited altogether. This concept took root after the Second World War and remains an important tool in the defence against relentless urban spread. In 2017, some 16,000 km^2 of England, 13 per cent of that country's land area, were designated as greenbelt. But greenbelt is far from sacrosanct. Local authorities recently implemented new plans that have already resulted in a loss of almost 800 ha of these 'protected' zones (Figure 4.1) (GOV.UK, 2017). Although this is only a small fraction of the total greenbelt zone, it is not a good omen for the future as housing pressures continue to mount.

Not all development is on greenfields, though, and there is continuous pressure to prioritise brownfield sites for new builds. These are areas previously used for purposes such as factories or waste dumps, but which are now derelict and available for reinvestment. There are probably about 35,000 ha of brownfield land in England and reutilising this resource to meet housing needs is a sound idea but in practice has limitations. Developers are often reluctant to work on these sites because they are more expensive to prepare for building work than a grassy, untouched pasture or meadow. Moreover, long-standing brownfield sites, however unappealing aesthetically, can sometimes be surprisingly

Figure 4.1. A housing development in the English countryside.
Source: ChrisHepburn/E+/Getty Images

valuable for wildlife. Bees, wasps and hoverflies often abound on them and some rare species such as the shrill carder bee (*Bombus sylvarum*) now rely more or less entirely on brownfields. Canvey Wick in Essex, Britain's first brownfield nature reserve, is a startling example of how important these places can be for wildlife. Its complex mixture of habitats supports a wide range of plants, insects, birds and reptiles now largely excluded from intensively farmed agricultural land. Even if brownfield sites sufficed to accommodate Britain's ever expanding housing needs, which they do not, indiscriminate exploitation of these places would have seriously adverse consequences for some of our most needy wildlife.

4.2 Railways and Roads

A century ago, the UK was criss-crossed with a complex matrix of railway lines that provided access to most of the country. This situation changed dramatically in the mid-twentieth century as road transport came centre stage, and line after line closed following recommendations of the Beeching report in 1963. About a third of the rail network was lost, including a track servicing our Somerset village. Railways in the past probably had minimal deleterious effects on wildlife and some of both existing and derelict cuttings and embankments continue to provide refuges for plants and animals disenfranchised in modern farmland. Very likely, these trackways facilitated the invasion of towns and cities by many of today's urban wildlife, especially foxes. The future, however, looks less benign. Railways are becoming fashionable again and high-speed intercity links, in particular, have become popular aspirations of politicians and industrialists. The first of these, HS1, running from London to the Channel tunnel, progressed with relatively little permanent damage to wildlife, but HS2 threatens a very different outcome. The first phase of this development, from London to Birmingham, will, according to reports from the Wildlife Trusts, threaten five wildlife refuges of international importance, 33 SSSIs, 693 Local Wildlife Sites and 21 local nature reserves. Thirty-two areas of ancient woodland face substantial damage and a further 19 will be completely destroyed if a second phase with links to northern England goes ahead. Political acumen and ecological know-how with respect to HS2 were typified by an environment minister who seemed to believe that ancient woodlands could be moved, or simply replanted somewhere else.

There were currently (in 2019) more than 246,000 km of roads, of all sizes, in the UK. About 189,000 km of this network is in England, which has about 1.45 km of roads per km^2 of land surface. As with urban sprawl, this puts England at the high end of European road densities, surpassed only by Belgium and the Netherlands. The total length of the UK road network increased between 1960 and 2019 by more than 27 per cent, amounting to some 85,000 km. There has been a corresponding increase in traffic levels, from 410 billion km travelled altogether in 1993 to 570 billion km in 2019, an increase of 39 per cent. Around 80 per cent of these journeys were made by cars and taxis (https://roadtraffic.dft.gov.uk/summary). It is almost impossible to relate meaningfully to such large numbers, but from common experience we know that all too frequently they translate into interminable traffic jams that have not noticeably declined commensurate with expansion of the road network. From the wildlife perspective, roads constitute immediate danger to life for any animal attempting to cross them, but they also pose a serious threat to habitat continuity with consequences including population fragmentation and genetic deterioration. Road construction inevitably destroys habitats directly, sometimes including precious sites with rare species that are apparently considered expendable by planners. My first encounters with wild reptiles in Britain came about on a glorious stretch of lowland heath, named after the village of Bagshot, in Surrey, where I lived at the time. A wide, shallow valley adorned with purple heathers surrounded a shallow bog pool and the place abounded with viviparous lizards (*Zootoca vivipara*), slow worms (*Anguis fragilis*) and adders (*Vipera berus*). Unfortunately, this unprotected piece of 'wasteland' caught the eyes of motorway planners, and most of it disappeared under the M3 in the early 1970s. Only later was it discovered that, based on nearby sightings, this piece of heathland probably also supported rare smooth snakes (*Coronella austriaca*). Driving that road now invariably engenders memories both pleasurable and sad.

4.3 The Advent of Statutory Protection for Wildlife in the UK

By the middle of the twentieth century, it was apparent that the British countryside with its plant and animal communities was under increasing threat from a whole range of impending developments. Protection at that time was minimal. The RSPB initiated a series of nature reserves in the 1930s, but most of the landscape was up for grabs, and the 1930s also witnessed a major housing boom. Inevitably, there was more to follow. Fortunately, some far-sighted individuals emerged at the end of the Second World War intent on preventing or at least arresting unregulated damage to the countryside. This was a time when public works for the common good were popular and high-profile reforms of education and creation of the National Health Service were set to revolutionise the quality of life in post-war Britain. Arthur Tansley, an eminent botanist and pioneer of ecological research, was prominent among those stepping up to the plate to attempt a comparable improvement in the prospects for nature conservation in the UK. Tansley (1946) recognised that some developments were inevitable, even desirable, but that 'progressive urbanisation and suburbanisation of the countryside inevitably destroy much beauty and diminish the rural area'. He went on to stress that 'the character and beauty of our countryside are very seriously threatened by impending developments. . .[by] unrestricted developments of the sort that is foreshadowed in many of the schemes for the future'. Tansley also anticipated difficulties promoting nature conservation: 'There is a school of thought which altogether undervalues the vital importance of our varied scenery and wild nature in the life of the nation.' He quotes an unattributed source espousing that 'Britain is an industrial nation. The important national tasks now are to secure industrial and economic prosperity. . .if in doing these things we sacrifice most of what is left of the former rural scene, what is that in comparison with the primary human importance of making a better society?' Sadly, this ill-informed misunderstanding of human needs persists among some industrialists and planners half a century on. Conflicts between developers and conservationists are unlikely to vanish anytime soon.

Fortunately, the recommendations of Tansley and his colleagues found favour with the government of the day. A new statutory body, the Nature Conservancy, was created to initiate and oversee countryside protection measures. The National Parks and Access to the Countryside Act of 1949 was the seminal piece of legislation that underpinned the ambitious conservation programmes that quickly followed. Levels of wild land protection were established, including National Parks, National Nature Reserves (NNRs) and the clumsily annotated SSSIs (or Areas of Special Scientific Interest, ASSIs, in Northern Ireland). The parks were primarily designated for recreation in places of scenic beauty, while nature reserves and SSSIs were more numerous but smaller and focused on wildlife conservation. In 2020, there were 224 NNRs in England, about one-third of which were managed by approved organisations (such as the Wildlife Trusts) and the rest by Natural England, the latest morph of the original Nature Conservancy. These reserves constituted about 0.7 per cent of the English land surface. In addition, Scotland had 49 NNRs, Wales 76 and Northern Ireland 47. Many other organisations have added their own nature reserves to the UK list, including the Wildlife Trusts and the RSPB, which has 200 of them. SSSIs are generally smaller than NNRs, although many NNRs are also wholly or partially classified as SSSIs. England has around 4,000 SSSIs, some of which have had their protection level strengthened by additional designations as SACs, Special Protection Areas (SPAs) or Ramsar sites, and collectively take up about 7 per cent of the landscape. In Scotland, there are over 1,400 SSSIs amounting to 12.6 per cent of the country's land surface, while in Wales some 1,000 SSSIs cover more than 12 per cent of the country, and Northern Ireland has more than 80 ASSIs. Greenbelts were established as mentioned above, and later came Areas of Outstanding Natural Beauty (AONBs).

Since that time, there have been further legislative provisions, which have strengthened levels of wildlife protection and included transfers of responsibilities to devolved governments in the four nations of the UK. Less agreeable has been a progressive dismembering of the original Nature Conservancy via a series of name changes, fragmentation into devolved organisations and an overall, continuous reduction in resources. Diminution of this devoted and effective band of conservationists would surely have disappointed Tansley, who died in 1955 at a time when there was cause for optimism about the prospects for British wildlife. Although the legacy of the 1949 Act and its various successors is in many ways still positive, the fate of Britain's wild places and their living entourage has become a major cause for concern. The seemingly impressive list of protection measures for ecologically important sites listed above include less than 10 per cent of the UK's total land area, leaving more than 90 per cent completely open to the overtures of developers, and even the nominally protected areas have not always remained sacrosanct.

The spread of suburbia and its associated infrastructure is the most obvious indicator of people pressure on the British countryside. Nevertheless, the question arises as to how much of the nation's wildlife declines can be attributed to this cause relative to other pressures that have also increased since the Second World War.

4.4 Impacts of Urbanisation on the Countryside and its Wildlife

The levels of habitat and species protection legislated for over the past 70 years have not proved sufficient to stem damaging urban encroachment and associated wildlife declines. Allegedly protected areas have been eroded in both quantity and quality over this time, and by the turn of the twenty-first century, only 46 per cent of SSSIs were judged to be in favourable condition for wildlife. Some had been damaged and others completely destroyed. Analysing the patterns of SSSI deterioration indicated little in the way of common features, but there was a weak positive association of damage with human population density (Jackson & Gaston, 2008). The downward trend has continued, and by 2010, just 37 per cent of SSSIs were judged to be in a good state. Reasons for this disappointing situation are multiple, but urbanisation and expanding leisure facilities have certainly played a role in wildlife declines both within and outwith protected areas.

Impacts of Urbanisation on Widespread Wildlife

An investigation into local declines and extinctions of rare plants in Britain made no assumptions about protected versus unprotected sites, but showed that since 1970, local extinction rates correlated with human population density (Thompson & Jones, 1999). Evidently, agricultural intensification was not, as widely assumed, the sole major cause of plant declines, and urbanisation has contributed significantly to losses of rare plants in the UK countryside. Many bird species are also suffering from the consequences of expanding developments, a problem set to intensify as built-up areas expand further (Gillings, 2019). For both breeding and overwintering birds in lowland England and Wales, the abundances of 95 per cent of 146 species were related in some way to human population density, which was used as a surrogate measure of housing extent. Although there was much variation among species, farmland birds such as skylarks were among the most negatively affected by urbanisation, while familiar garden birds such as magpies (*Pica pica*) and some wetland species showed an opposite trend, possibly due to creation of ponds and lakes in recreational green spaces. Low-density housing was generally beneficial to most birds, but at high human densities, up to 75 per cent of species were negatively impacted. Yet there have been some pleasant surprises in bird occupation of urban habitats, foremost among which has been the improved status of peregrine falcons (*Falco peregrinus*). These legendary predators are now tourist attractions in cities across the country, where they have taken to nesting on high-rise buildings and feasting on abundant urban pigeons. Recovering from a pesticide-induced population crash in the mid-twentieth century, these city falcons now experience greater nesting success and produce more fledglings than their rural relatives (Kettel et al., 2019). This is a welcome but rare conservation success story, albeit one that came about spontaneously and without human promotion. Sadly, it is an unusual outcome, as more often than not the spread of urban sprawl has been detrimental to wildlife across many taxonomic groups. Among the invertebrates, butterflies have certainly taken a hit from urban expansion. Comparisons of systematic butterfly counts in 442 urban and 3702 rural areas, mostly in England, showed that declines of 28 species monitored between 1995 and 2014 were commonplace everywhere. However, for 25 of these popular insects, the downward population trends were sharpest in urban zones (Dennis et al., 2017). The worst-affected species included large skippers (*Ochlodes sylvanus*), wall browns (*Lasiommata megera*), peacocks (*Aglais io*) and small tortoiseshells (*Aglais urticae*). All of this is disappointing news for those of us hoping that gardens will remain a valuable sanctuary for wildlife increasingly excluded from the farmed countryside. Butterflies are a well-recorded group of insects and the suspicion arises that their depletion in urban settings may well be mirrored by other genera for which little or no information is available.

Urbanisation Damage to Rare Habitats and Species

Of special concern has been the expansion of urban land into areas of rare and special habitats, sometimes despite the existence of legal protection. Sand dunes are one such habitat, and dune systems, among the most pristine landscapes remaining in the UK, can be found at many localities around the British coastline. West-facing shores, braced against prevailing Atlantic winds, are particularly well endowed with dunes. Those extending 20 km along the Sefton coast between Liverpool and Southport constitute one of the nation's largest dune systems, covering 2,100 ha, and are also one of the most biodiverse (Smith, 1999). They are home to an astonishing array of plants and animals, many of which are specialised to live predominantly in sandy habitats and some of which are nationally rare. Outstanding plants include sharp club rush (*Schoenoplectus pungens*) and petalwort (*Petalophyllum ralfsii*), among others. There are many interesting invertebrates; at least 16 species of dragonflies occur as well as 23 butterfly species, and beetles such as the spectacular tiger beetle (*Cicindela hybrida*). Two vertebrates of national conservation concern, the sand lizard (*Lacerta agilis*) and natterjack toad (Figure 4.2), also live there and the Sefton dunes support the largest surviving UK population of this rare toad. The national importance of this area was recognised early on by the Nature Conservancy, and was partly protected by the establishment of a NNR and several SSSIs in the 1960s. This was not sufficient,

Figure 4.2. Natterjack toad. (A black and white version of this figure will appear in some formats. For the colour version, please refer to the plate section.)

however, to prevent planning approval for a substantial new housing estate and a holiday camp on the dunes, the latter within a SSSI, at the end of that decade. I was among those collecting natterjack toads, by the dozen, in front of the advancing bulldozers and releasing them elsewhere on the dune system. Merseyside is one of Britain's most densely populated regions, and pressure for increased housing had proved irresistible, even on this precious habitat. Better news is that this scandalous desecration, as it was widely reported to be, led to a more enlightened attitude by the local authority, and since that time the remaining dunes have been comprehensively protected by further networks of SSSIs and other nature reserves. No more land there has been lost to development, although unfortunately much of the dune system has deteriorated for other reasons, mainly due to natural succession into scrub woodland.

Sefton is one of all too many instances where legal protection has been successfully overruled by development pressures. Lowland heath is another precious but rare habitat, confined mainly to southern and eastern England and home to a range of specialised plants and animals. Undulating hills and valleys decorated with heathers and gorse, purple carpets in late summer and autumn, are homes to species rarely or never found in other places. Go to the southern heaths of the Weald or, especially, Dorset to see sand lizards, smooth snakes and Dartford warblers (*Curruca undata*). Created by humans following forest clearances starting several thousand years ago, heathland has a wild, untamed character that appeals to naturalists and developers for very different reasons. Historically, heaths were frequently described as barren wastelands ripe for improvement, which has been widely interpreted as ideal places to build housing estates. Dorset is especially well known for its extensive heaths, but today they are a mere fraction of their former glory. More than 80 per cent of the county's heathlands have been destroyed one way or another since the start of the nineteenth century. As with the Sefton dunes, the national importance of the Dorset heaths was recognised in the early days of nature conservation, and several NNRs and SSSIs were created to defend what was left of them. Despite these efforts, housing developments have been a major and ongoing cause of heathland losses. In the mid-nineteenth century, Bournemouth was a small coastal village set in a sparsely populated and vast area of common land. By 2020, the city's urban sprawl was home to over half a million people. Inevitably, this growth needed ever more houses, which meant encroachment on the surrounding countryside. Heathland, despite its acknowledged value for wildlife, was still considered free range for developers. However, Canford Heath, near the adjacent town of Poole, became a cause celèbre for conservationists at the start of the 1990s. This is one of the largest surviving blocks of heathland in Britain, but up to and including the 1980s, piecemeal development had chipped away at it. In 1984, plans were laid to build over the whole area but were put in abeyance when, in response to this threat, the heath was belatedly scheduled as a SSSI. In 1991, the Secretary of State for the Environment made an unusual and dramatic intervention, revoking planning permission for the entire area. This turned out to be a significant event in the history of wildlife conservation, with most of Dorset's remaining heathland better safeguarded in subsequent years, but it was also a reminder that the needs of an increasing population for adequate housing have not receded.

Despite the legal strengthening of SSSI protection, sites supposedly safeguarded still come under threat. Nightingales rank as one of Britain's best-known birds (Figure 4.3). Most people have probably never seen or heard one, and they certainly never sang in Berkeley Square, but their reputation as charismatic crooners goes before them. Sadly, nightingales

Figure 4.3. A nightingale singing in scrub. *Source*: Roger Tidman/Corbis Documentary/Getty Images

are in steep decline, with fewer than 10 per cent of the number visiting our shores in the 1960s still arriving here in the twenty-first century. All the more important to safeguard what is left, and Lodge Hill in Kent is the best site in Britain for this summer migrant. Like most other birds, the nightingale is a protected species and surely its most important remaining piece of habitat should be safe from development. The local council disagreed, and approved the construction of 5,000 new houses there in 2014. In another belated rearguard action, Lodge Hill was scheduled as a SSSI in an attempt to avert its destruction, an action that allegedly annoyed ministers keen to support the housing plans. The original developer then pulled out, but another planning proposal swiftly superseded it. Finally, in 2018, there was agreement that just 500 houses would be constructed, all outside the protected area. Even so, the RSPB has expressed concern that housing will end up so close to the SSSI that associated damage will still have an adverse effect on the nightingale population. This important wildlife site is still by no means completely safe.

SSSIs are increasingly recognised as valuable national assets but developers still press their noses against the fence, and sometimes get their way. In an astonishing decision, the Scottish Government decided that Foveran dunes in Aberdeenshire could be developed as a golf course by American businessman Donald Trump. After much local controversy including attempted evictions of people living there, the course opened in 2012. The land was originally scheduled as a SSSI because of its outstanding character as a dynamic, mobile dune system. Its conversion for golf involved the construction of tracks, fairways and greens, as well as stabilisation by the planting of marram grass. The dynamic nature of the dune system, arguably its most important characteristic, was therefore irrevocably damaged. After several years, it became clear that the dunes had been altered beyond repair, and the very unusual decision was taken to remove its SSSI status. The decision to permit development was based on a judgement that economic benefits outweighed environmental considerations, but less than 100 of the 6,000 permanent jobs promised by the developers actually materialised. This was not an example of urban development pressure for more housing winning out, but a victory for a much less worthy assault by the leisure industry on our wildlife heritage. How dispiriting it is that politicians still accept economic arguments, however ill-founded, over the need to safeguard our wildlife heritage for future generations.

Gardens and Wildlife

Gardens have, for some time, been identified as safe havens for wildlife displaced from an increasingly hostile agricultural scene. Among those suffering from the widespread destruction of rural habitats, amphibians were among the first animals to be noticed benefitting from garden space. The newspaper headline 'Where have all the frogs gone?' reflected a recognition of damage done by agricultural changes as early as the late 1960s, but was increasingly answered with the repost, 'In my back yard.' More recently, urban habitats have become an important sanctuary for hedgehogs, those endearing small mammals that have declined precipitously since the 1970s. Like many others, we have managed our garden to promote as wide a range of wild plants and animals as possible. Cowslips, ragwort, violets, bluebells, songbirds, wasps, bees, hoverflies, butterflies, moths, beetles and many more seem to flourish, while ponds, those most marvellous of garden attributes, support newts, frogs, toads, water bugs and beetles, as well as damselflies and dragonflies. Gardens cover at least as much ground altogether as the county of Suffolk, and more housing estates may therefore have a plus side if they are accompanied by extensive outdoor plots. But how useful have gardens actually been in the overall scheme of things?

Widespread surveys across the UK have confirmed the popularity of gardening with wildlife in mind. Davies et al. (2009) found that almost 23 million households, about 87 per cent of all homes, had gardens, more than 12 million of which provided food for wild birds, and over 7 million also had nest boxes. There were between 2.5 and 3.5 million garden ponds at that time, and almost 29 million trees. This is an impressive tally. Less positively, recent studies have shown that most gardeners mow their lawns at least weekly and more than half continue to use chemical fertilisers or pesticides, practices not ideal for many members of the garden wildlife community. Furthermore, human population density has negatively influenced the number of wildlife-friendly features in gardens as well as their bird diversity, partly because newer housing developments usually have smaller gardens than many older dwellings. But whatever the caveats, bird feeding in particular is evidently a good thing to do. Following a remarkable calculation (Orros & Fellowes, 2015), it was reckoned that garden provisioning across the UK is sufficient to support at least 31 million 'hypothetical average' garden bird visitors. This amounts to almost half of the total UK populations of the 10 species included in the study. Knowing its likely value is especially comforting while watching goldfinches (*Carduelis carduelis*) and house sparrows (*Passer domesticus*) squabbling in droves around our feeders, a more or less continuous daytime activity at all times of year. But despite this pleasing news, most bird species fare less well in urbanised environments than elsewhere, as described earlier, and the same is true of butterflies. Gardens also often support high densities of predators, especially domestic cats (*Felis catus*). There is no doubt that cats impact adversely on garden wildlife, especially birds and small mammals, which they kill in substantial numbers. Despite this carnage, their overall effects on populations of their victims remain debatable, but there can be sublethal as well as direct consequences of prowling cats, which, in the case of blackbirds (*Turdus merula*), involve increased nest predation by opportunists such as grey squirrels (Bonnington et al., 2013). RSPB surveys recently revealed that sightings of frogs, those early beneficiaries of garden ponds, declined by more than 40 per cent in gardens between 2014 and 2019. There are probably fewer ponds in newer, small gardens, but the frog issue may also relate to the fragile ecology of garden habitats. In many instances, a successional process occurs in which early proliferation of frogs is followed by newt colonisation. After a few years, large numbers newts of build up and devour frog spawn so effectively as to eventually exterminate the larger amphibian. I have twice witnessed this progression in our ponds. Garden ponds are usually much smaller than the farm pools they have replaced, and the confined space probably facilities this destructive competition to an extent less likely to occur in larger habitats with lower amphibian densities. Scale matters, and the inherently small areas of gardens cannot adequately reproduce the ecology of large rural expanses. The evidence shows that while garden are certainly valuable refugia for many wildlife species, these benefits are small compared with the overall destructive impacts of urbanisation.

4.5 Roads and Wildlife

Britain's road network in the twenty-first century is pervasive across the entire nation, greatly exceeding what the railways achieved in their nineteenth-century heyday. Few places, mostly in the uplands, are now more than a few kilometres from a highway. But this connectivity has come at a high cost to wildlife. Roads pose three particular threats. Their very construction often involves the destruction of places important for wild plant and animal communities. After that, road traffic becomes a killing spree, accounting for huge numbers of animal deaths every year. And roads partition the landscape, fragmenting wildlife populations and thus rendering them liable to the damaging consequences of reduced connectivity. These effects are serious and all have impacted on the state of Britain's wildlife.

Road Construction and Its Aftermath

In 1994, the same motorway that wrecked my patch of Bagshot Heath continued on its destructive path, this time through part of the Twyford Down SSSI in Hampshire. This area of outstanding chalk grassland, a rare and precious habitat, was home to a range of interesting plants and invertebrates including six orchids and chalkhill blue butterflies (*Lysandra coridon*). This time, the M3 did not get its way unchallenged, and construction was subjected to extensive protests by a range of objectors including the local residents, the so-called 'Dongas Tribe', New Age travellers and women from the Greenham Common peace camp (Figure 4.4). But direct action and legal appeals came to nought, the supposed safeguard of SSSI status was overruled and the motorway continued on its way to the south coast. This is but one of many examples of how road building has regularly taken precedence over nature conservation. Another startlingly bad example came only a couple

Figure 4.4. Some of the many objectors at Twyford Down. *Source*: Stuart Freedman/Corbis Historical/Getty Images

of years later, with the creation of a bypass for Newbury in Berkshire. This six-lane highway destroyed healthland, bogs and wildflower meadows and damaged the River Kennet, one of England's most beautiful chalk streams. Archaeological sites, nature reserves, an AONB and three SSSIs went under the bulldozers, despite more mass protests and legal challenges. At particular risk was the rare Desmoulin's whorl snail (*Vertigo moulinsiana*), a small inhabitant of swamp litter. Again, the development proceeded unchecked, and mitigation measures to move snails to an alternative habitat ultimately failed to generate a new population. These well-publicised examples are, unfortunately, the tip of a large ice (or rather tarmac) berg. In the 1990s, road plans threatened more than 190 SSSIs in England alone. Although levels of road building fluctuate over time as a function of government whim, they never cease, and extensive new programmes were announced in 2020. At least most new road schemes invading protected sites are now usually accompanied by some form of mitigation, and this can work reasonably well. At Twyford Down, the M3 replaced the old Winchester bypass between Bar End and Compton, which was then closed and removed, and planted with species-rich grassland. Mitigation, however, is a chancy business, as demonstrated by the sad fate of whorl snails at Newbury. With ever more motorists on Britain's roads, it is hard to see how this pressure to expand the network, whatever the cost to wildlife, will ameliorate.

Road Kill

We know that roads are dangerous places. In 2019, almost 1,900 people died in road accidents in the UK, dreadful tragedies for so many families. Better news is that road deaths have decreased by 75 per cent since 1970, despite ongoing increases in traffic volume, the improvement presumably a result of wide-ranging new safety measures. Unfortunately, this pleasing trend has not been reflected in road kills of wildlife, of which the numbers are huge and involve many different species. Sometimes human and wildlife disasters on roads combine, especially when deer are involved. An astonishing 74,000 accidents involving these hefty mammals occur annually, typically generating several hundred injuries to people and a score of human deaths. Although most of the deer involved die following these incidents, Britain's burgeoning deer populations will not be compromised by this level of mortality. However, most of the wildlife decimated on Britain's road network are small species with which collisions often go unnoticed and are invariably fatal.

In Wales, all species of ground-dwelling vertebrates occurring in that country have been found as corpses on roads, as have some bats and birds (Slater, 1994). Indeed, probably every species of British mammal has been killed somewhere on our highways, and the national list includes at least one (non-native) wallaby and a recently reintroduced beaver. There is a vicious circle in which initial casualties, mostly nocturnal, are scavenged by latecomers, some of which then also fall victim to traffic. Quantifying road kill is difficult for several reasons. Numbers vary according to whether they are counted on foot, on a bicycle or by car, and, unsurprisingly the first of these methods invariably yields the highest scores. The tolls also regularly change according to season and even by day of the week. Effective scavenging can make a mockery of any counting method, especially if this is delayed substantially after dawn. At one location, 179 toad corpses were noted at daybreak, but within a couple of hours every one had been removed by opportunistic scavengers. This implies that road mortality can be assessed most reliably for species too large to be removed quickly by other animals, and in the UK this means mainly badgers, foxes, otters, hedgehogs and owls. Hedgehogs have long been the most commonly noticed dead animal on Britain's roads, and surveys have suggested that between 167,000 and 335,000 of them perish under our wheels every year, despite a sharp national decline in this popular creature. At the start of the twenty-first century, surveys indicated that hedgehogs constituted 29 per cent of mammalian road casualties, badgers 25 per cent and foxes 19 per cent. Among the birds, 25 per cent of the dead were tawny owls (*Strix aluco*), 19 per cent were kestrels (*Falco tinnunculus*) and 16 per cent were barn owls (*Typo alba*). These owls, fluttering along roadside verges, are particularly vulnerable to collisions with passing vehicles, and it may be that they cannot now sustain viable breeding populations close to trunk roads and motorways. Road casualties may be responsible for suppressing the populations of common mammals on a local scale. Some of these statistics are changing. According to Project Splatter (https://projectsplatter.co.uk), a citizen science study based at Cardiff University, badgers and pheasants have become increasingly dominant in road casualty counts. This certainly tallies with my experience in recent years, during which time I've seen precious few dead hedgehogs but all too many deceased badgers. This carnage is a sorry sight, and it is a rare road trip that doesn't encounter at least a corpse or two. For a few individuals, though, this experience triggers altogether different emotions. Collecting dead animals from roads is legal so long as the creatures were killed by somebody else, and there are those who make a meal of what they find. One couple, rather than cooking turkey or goose at Christmas, revels in a leg of roadkill. Scouring local roads apparently produces almost 50 edible corpses every year for them to devour, including deer, pheasants, rabbit and badgers. The habit has not proved universally popular with friends and family, however, some of whom describe the enthusiasts as 'serial killers'.

An important and inevitable question is whether these ongoing massacres generate population declines in the susceptible species. Hedgehogs are a case in point. Evidence suggests that 'Tiggy-winkle' numbers in the UK have crashed from perhaps 30 million in the 1950s to around 1 million in the 2010s. The numbers of dead hedgehogs reported on roads has declined commensurate with the national population size, but road deaths are not considered to be a primary cause of problems facing this prickly mammal. Intensification of farming methods are thought to the main culprit, a conclusion supported by the relative success of hedgehogs in suburban gardens. Declining numbers of road corpses, on this basis, simply reflect fewer animals around to attempt crossings. Badger road mortality tends to support this explanation, as badger road kill has increased in line with overall increases in badger numbers over recent decades. However, for another prominent group of road casualties, the picture may be different. Mortality of amphibians migrating across roads to their breeding pond has been a cause of concern for several decades. Despite the difficulties with corpse counting discussed above, there is a widespread consensus that amphibians are the animals killed in the largest numbers on roads. Common toads (*Bufo bufo*), in particular, move in large migrations over short peak periods in spring, and corpses may be so plentiful that they become a skid hazard for passing cars. An early response to this horror story was identification of places in the UK where the carnage was especially severe, and the appearance of volunteer groups taking their lives in their hands by carrying toads across the highway. In this way, a Toads on Roads campaign got going in the 1980s and has since included over 600 specific localities. On busy roads, the probability of an amphibian being killed attempting to cross unassisted approaches 1, and for some species up to 10 per cent of the population can perish annually on their springtime treks. Common toads have declined continuously since at least the 1980s in Britain and in some European countries (Petrovan & Schmidt, 2016), almost to the extent of qualifying for IUCN red listing status. To what extent is this due to road mortality? Connecting road deaths to population trends is fraught with difficulty, but there is some indicative evidence. In an area where surrounding habitat was largely unchanged, Cooke (2011) found that numbers of toads killed on roads en route to three separate ponds declined over 20 years contemporaneously with numbers in the ponds and negatively correlated with increasing traffic flows (Figure 4.5). The Toads on Roads volunteer efforts failed to halt this trend. Several studies in Europe and North America have also shown that roads can cause amphibian declines, detected directly or by comparing variations in species diversity as a function of distances from highways (Beebee, 2013). There is often a lag period of several years, or even decades, before population declines are apparent as mortalities accumulate. Amphibians are unusual both by being particularly vulnerable to road deaths and also by it being easier to relate these deaths to population trends than is usually the case with other taxa. Roads can also be a source of lethal pollution. Adjacent ponds and ditches are susceptible to run-off of toxic chemicals, especially oil, but even salt applied for deicing in winter has caused substantial mortality of great crested newts attempting to migrate across a treated road surface.

Roads can affect animal abundance even when killing is not necessarily involved. Thirty out of 51 widespread birds in Britain varied in numbers according to proximity to roads, 20 of which were positively

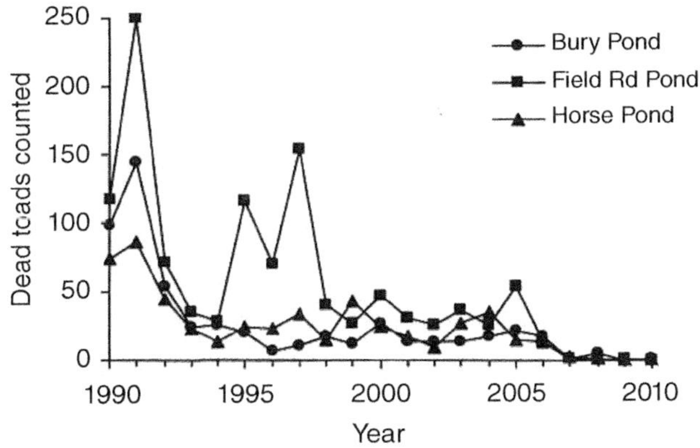

Figure 4.5. Declines of three toad populations associated with road traffic. © Arnold Cooke

correlated with minor roads (Cooke et al., 2020). These largely comprised species commonly associated with suburbia, such as swallows, chaffinches (*Fringilla coelebs*) and goldfinches. Major roads, in contrast, were often bad news, with blackbirds, chaffinches, house sparrows and starlings (*Sturnus vulgaris*) among those in which populations were negatively affected. It is surely remarkable that highly mobile animals that rarely alight on road surfaces nevertheless show population-level responses to modern highways.

Roads and Habitat Fragmentation

A less obvious but invidious effect of roads is their fragmentation of habitats and of species populations on either side of them. Before dwelling on this dark side of what roads can do, it is worth noting that in some circumstances they can function positively as wildlife corridors. Motorway embankments, out of bounds to the general public, sometimes support populations of plants and animals more at home there than in surrounding overmanaged countryside. As linear features, they also permit wildlife to move over considerable distances, a property that might become increasingly valuable where species need to alter their distributions in response to climate change. One scheme in south-west England included planting 10,000 native trees and shrubs to fill or reduce gaps in hedgerows and woodlands beside the A30 and A38, thereby connecting over 105 miles of habitat on the verges and on land adjacent to the two roads. In addition, almost 27,000 m^2 of new heathland were created along the edges of roads in Dartmoor in Devon, and Bodmin Moor and Goss Moor in Cornwall. On a smaller scale but for similar reasons, the verges of minor roads can provide valuable habitats, especially for wildflowers, although this is too often compromised by overenthusiastic management. An obsession with neatness comparable with gardeners and tidy lawns has meant that local authorities frequently cut back roadside vegetation to within centimetres of its life. There has been a 20 per cent decrease in floral diversity on road verges since 1990, in part because of overzealous cutting. This needs to change, and even the potential benefits of road connectivity can sometimes have a downside. In Australia, invasive cane toads (*Rhinella marina*), an ongoing threat to many native species, have used roads as conduits to expand their range faster than they manage in other habitats.

Fragmentation of previously contiguous blocks of habitat is increasingly recognised as an invidious danger to wildlife. The landscape of the UK has changed markedly during the last 60 years with the contraction and fragmentation of semi-natural grasslands, upland and lowland heaths, freshwater wetlands and coastal margin habitats. About 77 per cent of SSSIs and 98 per cent of Local Wildlife Sites are smaller than 100 ha. For several Biodiversity Action Plan priority habitats, the median patch size remaining is smaller than 2 ha. The Dorset heathlands provide a good example of the extent of fragmentation that has occurred over the last 200 years at the landscape scale. In the early 1800s, heathland covered most of south-east Dorset. Some 86 per cent of Dorset's heathland has been lost since then, and the surviving area is broken into more than 150 small sites with lots of 'edge' and little 'interior' habitat. In this situation, road networks, especially when as dense as those in England, constitute potential barriers to individual movement, which can then disrupt connectivity among previously large populations. Even minor roads may deter movement of animals across them, as shown for voles and carabid beetles in Germany several decades ago (Mader, 1984). In most situations, there is no easy way to assess population fragmentation directly, but genetic analyses have provided powerful indirect methods to address the issue. Using highly variable molecular markers such as DNA microsatellites, it is possible to sample genetic diversity across a landscape and detect whether populations are differentiated genetically by barriers to gene flow such as roads. Comparing results of genetic studies on a range of insects, amphibians and mammals showed that differentiation across roads was widespread but not universal (Holderegger & Di Giulio,

2010). Roads were often associated with decreases in genetic diversity, the extent of which varied according to road density. The effective population sizes of two species, the carabid beetle (*Carabus violaceus*) and red deer were reduced in roadside populations, inferring increased susceptibility to inbreeding and genetic drift. One problem with this approach is that many roads are relatively new, but it takes several generations for genetic changes to become clear at the population level. Fortunately, some genetic methods now permit measuring levels of individual movement in the short term. Studies of this kind usually show that roads are rarely complete barriers to animal movements across them but also infer that the numbers that manage it are probably too small to prevent genetic fragmentation of a population bisected by a moderately sized or large road. Genetic studies in suburbia have also highlighted how road matrices effectively partition amphibian populations and can lead to local inbreeding issues with common frogs and toads, highlighting another limitation on the value of gardens to wildlife in the long term. Physical abnormalities, slow larval growth rates and occasional albinism are commoner in garden compared with rural frog populations, and reduced fitness may constitute another reason for recent frog declines in garden habitats.

Mitigation of the Road Effect on Wildlife

The dreadful carnage inflicted daily on so many wild animals is an indictment of the human devotion to travel anywhere at any cost. From the animal welfare point of view alone, every effort should be made to minimise road casualties, and the tireless work of volunteers carrying amphibians across roads on dark spring evenings exemplifies this admirable mindset. Unfortunately, conservation issues are not so readily addressed. Usually only amphibians can be helped in this way, and then just in one direction; animals returning later, after breeding, move in a more piecemeal fashion and cannot be saved. There is little evidence that 'toad patrols' ever avert long-term declines.

By far the commonest approach to ameliorating wildlife deaths on roads has been the construction of under-road tunnels. These can be wide enough to allow the passage of large mammals, but most have been narrower and created primarily for amphibians and other relatively small creatures. Tunnels often succeed in attracting animals to use them, but with such a vast road network, they are inevitably feasible at only a small number of sites. Apart from being expensive to install under existing roads they require, to be effective, long stretches of climb-proof fences extending in both directions from a tunnel entrance on both sides of the road. And there are maintenance requirements, especially prevention of fences being overgrown with vegetation that reduces their efficiency as barriers. Unfortunately, some such tunnels, including one I knew well in Sussex, do not fulfil all of these rather demanding requirements. But do at least some of them work well enough to prevent local population declines? From observations in several countries over many years, it has become clear that many tunnels fail in their objectives for various reasons, including lack of some of the requirements listed above. However, the good news is that if well designed and properly monitored, they can succeed. A study in Yorkshire demonstrated that four species of amphibians have used a series of tunnels and that numbers of the strictly protected great crested newt increased in nearby ponds over the 4 years of study (Jarvis et al., 2019). These tunnels were also used by other species including invertebrates and mammals such as brown rats, field voles (*Microtus agrestis*), wood mice (*Apodemus sylvaticus*), common shrews (*Sorex araneus*), hedgehogs and even otters. Until fairly recently, tunnels were the only effective mechanism for mitigating road casualties, but now there is also the option of creating a 'green bridge'. Construction of these wildlife corridors has increased in recent decades, especially in Europe but increasingly also in the UK, although at present there are very few in Britain (Natural England, 2015). Ideally, green bridges are at least 80 m wide and planted with vegetation typical of that occurring on adjacent sections of road. Some even include ponds, and as with tunnels, it is important to channel animals towards them with adjacent fencing. They have been used successfully by many species of amphibians, reptiles, mammals and invertebrates. Dormice (*Muscardinus avellanarius*) have been recorded on one of England's first green bridges in Kent, but there is much more information from the Netherlands, the European country with a particularly large number of these conduits. Six species of amphibians and eight of mammals, including deer, have taken advantage of Dutch green bridges to cross roads successfully. However, we have next to no information about whether usage levels are high enough to prevent population fragmentation. These conduits are very expensive to build, and like underpass tunnels, it will only ever be feasible to install them at relatively few sites. In most places, animal deaths are bound to continue unabated.

Evidently, roads have contributed to the declines of some species of Britain's wildlife and caused untold suffering to many more. With current and future expected levels of traffic transporting ever more people around the UK, this carnage can only get worse.

4.6 Pollution and Wildlife

Contamination of the environment with pollutants is another damaging human activity, downgrading the air we breathe, the water we drink and the seas in which we bathe. In this chapter, I exclude the contributions of agriculture to this problem, a topic that will be addressed in Chapter 5. Considering its impact on human health, it seems remarkable that, despite efforts over many decades to control it, serious damage from various types of pollution is ongoing. Every year, 36,000 people in Britain die as a consequence of air pollution, compared with fewer than 2,000 in road accidents. Almost all of the air pollutants are emitted in built-up areas or on busy roads. During the spring lockdown of human activity caused by the coronavirus pandemic, space-based sensors demonstrated dramatic reductions in air pollution over cities in many countries, including the UK. Predictably, this improvement did not survive relaxation of the lockdown. Living next to a city street can be a death sentence, and 80 per cent of the British population resides in urban environments. What, then, is the impact of pollution on wildlife?

Air Pollution and Wildlife

Undoubtedly, carbon dioxide (CO_2) is the most serious atmospheric pollutant globally and the main cause of climate change with its invidious consequences for life everywhere. This includes the UK, and the topic is considered as an issue in its own right in Chapter 6. Although CO_2 concentrations are at an all-time high, they are still low in relation to the other major atmospheric gases (nitrogen and oxygen) and do not constitute a toxicity threat to humans or wildlife. Unfortunately, the air in Britain is frequently contaminated with a range of much more dangerous chemicals, namely sulfur dioxide (SO_2), nitrogen oxides, butadiene, carbon monoxide, heavy metals, fine particles and toxic organic micropollutants. Diesel engines are a major cause of dangerous air pollution, and in 2009, the UK was described as one of the worst polluters in Europe for airborne particles and nitrogen dioxide (NO_2), while in 2007, it was condemned for exceeding EU limits on ambient concentrations of SO_2. These emissions from cars and industrial installations cause respiratory problems and aggravate cardiovascular disease. Efforts over the past decade have made some progress with improving air quality in Britain, but there is still a long way to go.

Wildlife, as well as people, has suffered from the chemical cocktails surrounding us. Impacts of air pollution on ecosystems have been wide-ranging and damaging (Paige Wright et al, 2018). The consequences of acidification, primarily by SO_2, have been well documented since the 1980s. These include increasing plant susceptibility to stress and mass mortalities of fish and amphibians. Emission controls have reduced acidification problems, but they have not disappeared entirely. More recently, nitrogen deposition has had a widespread eutrophication (nutrient enrichment) effect, including substantive changes in plant species communities. Both terrestrial and freshwater ecosystems have become impoverished by this ongoing process, which commonly reduces overall biodiversity and is especially severe near roads. Adverse effects of vehicle emissions on the ecology of roadside animals and, especially, plants have been widely investigated. Organisms unlucky enough to live close to roads regularly suffer from genetic abnormalities, increased mortality rates, impaired development and altered behaviour. At the community level, species composition can change in response to selection for toxin resistance or to increased eutrophication as a result of nitrogen oxide emissions from vehicle exhausts. Heathlands are inherently oligotrophic (nutrient-poor) environments and thus especially vulnerable to these pollution effects, and in the New Forest, changes up to 300 m from a main road included increases in vascular plants, especially grasses, and declines in lichens (Angold, 1997). Lichens are particularly susceptible to air pollution and in the late twentieth century suffered dramatic declines consequent upon high atmospheric SO_2 levels. Happily, concentrations of that particular gas have since been greatly reduced, but lichens now face a different problem due to the more recent increases in nitrogen gases, particularly NO_2. Lichen communities have therefore changed, usually with increases in nitrogen-tolerant species and declines in less tolerant ones. Nitrogen pollution is now also decreasing in many places, but there is no doubt that air pollution of various types has caused problems for a wide range of susceptible species. A particularly bizarre example came to light during a study of feral pigeons (*Columba livia*) in French cities (Jiguet et al., 2019). Toe mutilations, including complete losses, varied according to levels of air pollution and, of all things, the number of hairdressing salons. The authors speculated that loose hair cuttings might garrotte pigeon toes. Control of hairdresser activity seems an unlikely prospect, comfortable though some of us might be with such a policy.

On balance, impacts of air pollution on wildlife have been severe in the past, especially acidification, but some at least are on the decline. Roads, however, remain as a pervasive source of damaging emissions affecting humans and wildlife alike.

Freshwater Pollution and Wildlife

The British Isles are blessed with an enviable complement of freshwater ponds, lakes and rivers, as well as a lengthy and inspiring coastline bathed in bountiful seas. In pristine condition, these places are home to a huge variety of plants and animals, but many of them have proved susceptible to pollution. In the case of ponds and, to some degree also rivers, agricultural practices have been the main source of damaging contaminants, and this aspect of the problem is discussed in Chapter 5. Much of the harm inflicted on Britain's rivers and coastal waters, however, originates from high-density human settlements. Being at the forefront of the industrial revolution, with ever more people migrating into cities, Britain simultaneously gained an unenviable edge in generating formidable amounts of water pollution. Rivers were convenient conduits for free disposal of wastes emanating from factory effluents and from increasing sewage discharges. The well-known account of a stinking River Thames that finally stimulated politicians in the House of Commons to consider pollution problems has an unfortunate resonance to this day. The legislature has remained as resistant as ever to act on environmental issues, even when there is overwhelming evidence, as there is now, of a need for it.

The River Mersey is a classic example of an ecosystem devastated by pollution. As a youth, it was my closest waterway to home but not one any aspiring naturalist would be inclined to visit. In fact, it was rarely visible beneath a blanket of foam, with local folklore claiming that anyone falling in would die of toxic inhalation before they had time to drown. The horror story of this river's fate since the eighteenth century, with its astonishing cocktail of chemical and sewage discharges, has only recently included signs of recovery (Burton, 2003). Happily, improvements in the Mersey's water quality have continued and, after a nadir when it is believed no fish at all were present, it now supports a wide range of species including occasional salmon. Anglers can be seen on its banks once again, after more than a century of absence. That such a revival has been possible in one of the most densely populated river catchments in the UK is a notable achievement, but it remains the case that most of Britain's rivers are not in really good condition. There is a lot of political spin in the field of river quality assessments. Government departments are keen to point out how standards have improved in recent years, which is true. Conservation organisations more often highlight continuous failures to meet EU standards. A particular ongoing problem is contamination by storm water sewage, which regularly spills into rivers after heavy rain. Water companies have been slow to combat this pollution source. Recent tests showed that in 92 per cent of samples taken from the River Thames, coliform bacteria, probably originating in these overflows, were present. Although fish populations have recovered as rivers have become cleaner, other wildlife has not always been so fortunate. As recently as the 2000s, in suburban rivers, eggs of dippers (*Cinclus cinclus*), those delightfully bobbing birds of river torrents, contained some of the highest levels of organochlorine pesticides ever recorded in passerines. Despite their discontinued use, levels of these bioactive chemicals were stable or increasing and approached concentrations that can damage development (Morrissey et al., 2013). Although sewage treatment now reduces the amount of most organic waste entering rivers, some undesirable chemicals present in low concentrations are more difficult to contain and often still get through. Steroids, especially oestrogens, enter rivers this way and have caused feminisation in fish living downstream of the outflows (Figure 4.6). In some waterways, all of the male roach (*Rutilus rutilus*) examined showed signs of feminisation, including low-quality sperm and developing eggs in the testes. The consequences of altering the sex ratio on wild populations are not clear but are unlikely to be good. And ad hoc pollution incidents from accidental or deliberate releases of toxic effluents continue to haunt British rivers. As one of all too many examples, the River Frome in Dorset has suffered serious upsets, even in recent years. One pollutant turned the water a dramatic, luminous blue, while an earlier incident killed at least 1,700 fish. The great efforts to improve water quality have not gone unrewarded, but we are constantly reminded that wildlife in our beautiful river systems remains vulnerable and is likely to remain so as pressure from increasing human numbers intensifies.

Marine Pollution and Wildlife

Freshwater pollution is not constrained to remain in rivers, and much of it, especially sewage overflows, eventually finds its way into the sea. People enjoying a swim from an English beach still risk meeting contamination from this source, an ongoing concern for organisations such as

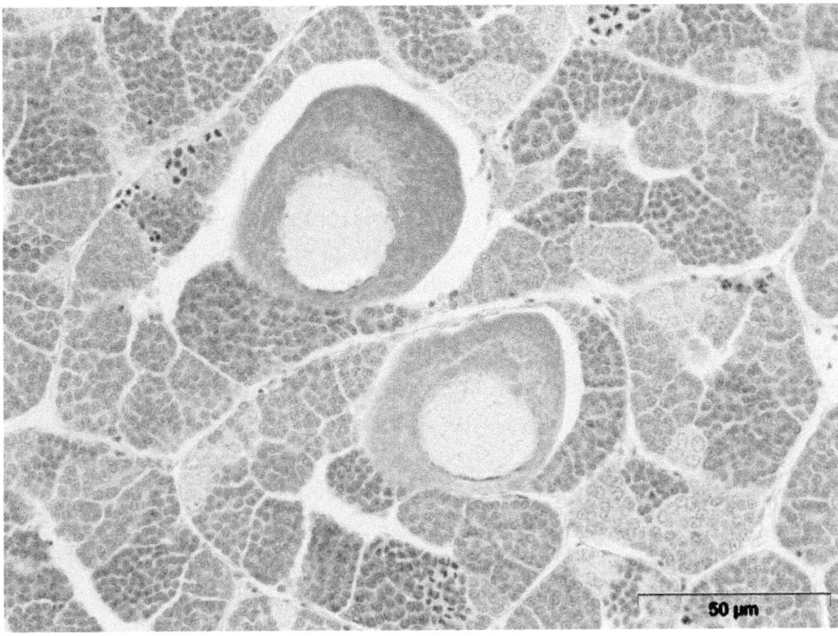

Figure 4.6. Histological section through the gonad of an intersex roach showing two primary oocytes developing within testicular tissue. Photograph provided by C. R. Tyler

Surfers Against Sewage. But the emphasis around British shores, as elsewhere around the world, has shifted to the impact of overwhelming plastic pollution. Quantities of this undegradable material have accumulated over decades to attain astronomic proportions. Marine plastic and non-plastic debris has been found in seabed sediments, the water column, floating on the sea surface and on foreshores. A walk along the strandline of almost any beach in the UK reveals the demoralising extent to which plastics have encrusted our otherwise glorious coastline.

Annually, up to 20 million tonnes of plastic waste end up in the world's oceans. Although not significantly degraded chemically, this material is physically pounded into ever tinier pieces invisible to the naked eye. These microplastic particles add an extra potential threat, largely unseen, to that posed by larger items. Some of these materials, both macro and micro, end up in animal digestive systems. Marine turtles are among those animals particularly prone to ingesting plastic, especially species that prey heavily on the jellyfish that such debris can simulate. In loggerhead turtles (*Caretta caretta*), models suggest that plastic ingestion reduces food abundance, body condition and reproductive potential, and that these consequences may lead to population declines (Marn et al., 2020). Entanglement is another issue that can have lethal consequences. Specimens of six marine turtle species have been found suffering from this danger, and of thousands of beach-stranded individuals seen every year, more than 5 per cent were entangled with discarded fishing nets and plastic debris, and more than 90 per cent of those affected animals died. Cetaceans are also highly vulnerable to death by plastic. Ingestion has been recorded in more than half of the world's cetacean species, inducing mortality in more than 20 per cent of stranded whales, dolphins and porpoises. There is evidence that 25 out of 69 species of north-eastern Atlantic seabirds have ingested plastic, and some species have incorporated it into their nests. Thus far, the effects of microplastics have mostly been studied in the laboratory. Under such circumstances, they can be ingested by zooplankton and from there pass ever higher in food chains at least as far as fish. Microplastics certainly occur in wild fish, and were detectable in more than a third of 500 fish caught in the English Channel. And microplastics are increasingly common in humans, detectable in several body tissues including the liver and lungs. Despite their increasing ubiquity, the effects of microplastics on the health of humans and wildlife, if any, remain to be discovered. What is clear, though, is that large plastic debris can kill marine animals by ingestion, essentially starving them, or by entanglement, which often has the same ultimate result. The scale of this mortality is now at a level where it could be precipitating declines of some turtle and mammal populations.

4.7 Overview

That building housing estates and ever more roads has impacted badly on wildlife is hardly surprising. Statutory protections legislated to safeguard the countryside have succeeded in part but have too often failed when development is given priority over wildlife, as has happened even on rare and precious habitats. Declines of birds and butterflies, among others, are associated with urban spread and pervasive road networks. Gardens managed for wildlife have had some mitigating effects, as have tunnels to facilitate animal movements under roads, but these have been small compared with the damage done. And pollution of air and fresh and sea water has accompanied these developments, further damaging wildlife in its wake.

The increasing amount of UK countryside under concrete and tarmac is clearly associated with the country's increasing human population. In 2015, there were 27 million households in Britain, a 7 per cent increase since 2005 and rising at a rate strongly correlated to human numbers. It has been suggested that increased demand for housing has been fuelled in part by more people choosing to live alone, but the proportion of households with single occupancy, around 29 per cent, has not changed significantly since 2005. The total number of cars on British roads has increased steadily since the 1950s, but the percentage of households with at least one car, around 77 per cent, was stable between 1995 and 2007. This apparent discrepancy is accounted for by continued increases in the average numbers of cars per household. A key statistic, however, is that the average annual distances driven per car declined by 20 per cent, from 9,200 to 7,400 miles, between 2002 and 2019. The ongoing increase in traffic across the UK, as described earlier, can therefore most reasonably be accounted for by increasing numbers of people behind the wheel. Sewage production is unequivocally proportional to the number of its producers, making it ever more difficult to contain pollution spills into rivers and coastal habitats. As for plastic, the main newcomer on the environmental degradation scene, we can only speculate that discards are likely correlated with people numbers, but of course an unknown proportion of the debris afflicting UK beaches hails from beyond our shores. The overall conclusion is clearly that increasing urbanisation, infrastructure and pollution have contributed to wildlife declines to an extent that relates directly to the UK's human population size.

5 · *Impacts of Farming and Forestry on Wildlife Declines*

5.1 The Great Outdoors

Beyond the garden gate, Britain's picturesque countryside spreads out before us. Hills and dales, woods and fields, bogs and moors, lakes and ponds, rivers and streams cover some 80 per cent of the UK's surface and beg to be explored. But all is not well in the land. When the post-war Nature Conservancy came into being, the threats to wildlife most quickly recognised were those of urban development and pollution, as described in Chapter 4. Farmers were 'guardians of the countryside' and the expectation was that farmland would continue to be a safe haven for British wildlife. This was a reasonable assumption. Farming suffered a depression in the 1930s, during which time much of the rural landscape experienced a level of neglect in which most wild plants and animals prospered. However, by the end of the Second World War, the writing was on the wall for a very different farming future. 'Dig for Victory' campaigns had destroyed much previously underutilised habitat, such as heaths and moors. More significantly, agrochemicals arrived on the scene including artificial fertilisers and pesticides, all dedicated to improving productivity and ensuring that the UK would become self-sufficient in food production. Governments had learnt lessons from the German U-boat blockade, but the costs to wildlife from its response would be high. By the early 1960s, these costs were beginning to emerge. Rachel Carson's seminal *Silent Spring* (1962) revealed the devastation being wrought by pesticides, especially DDT, that were killing non-target animals on a large scale in America and Europe. One of my earliest memories of that time was finding dead blackbirds grotesquely poised in hedges, victims of a poison that was eventually banned from general use. Unfortunately, lessons from Carson's ground-breaking work have not been learnt, and newer generations of pesticides continue to cause havoc with far too little regulation. Twenty years after the publication of *Silent Spring*, the state of Britain's countryside had deteriorated further, with chemically led intensive arable farming creating huge, hedge-free landscapes highly inhospitable to most wild plants and animals (Shoard, 1980). By this time, it was already a long shot to describe the farming community as countryside guardians, and instead industrial-scale agriculture has become the bête noir of wildlife conservation (Figure 5.1). And farming has extended beyond the shoreline, with intensive aquaculture and seabed trawling adding to the list of activities damaging wildlife and habitats in the UK's coastal waters. For many conservationists, farmers have become the villains of the piece.

Agriculture before the last war was relatively benign with respect to wildlife, but forestry, the other main human activity in the open countryside, had already embarked on a damaging path. Deforestation had left the UK with less woodland cover than any comparable European country, and after the First World War, the Forestry Commission was set up to remedy a timber shortfall. Its main response was to create large areas of conifer plantations, with serried ranks of close-packed trees that generated gloomy habitats inhospitable to most wildlife. This practice continued for many decades and only recently has progress been made in reversing its unfortunate legacy. Foresters as well as farmers have enjoyed low esteem with naturalists in recent decades.

In this chapter, escalating assaults on the British countryside and its wildlife in the post-war years are described, concluding with a consideration of how food and timber production have changed over that period. To what extent have the original goals been achieved, and how do these changes relate to wildlife decline and the UK's human population size?

5.2 Farming in Britain

A Post-War Agricultural Revolution

The countryside of the 1930s would have seemed, to those not living in it, like the rural idyll of picture-postcard England. For the farmers of that period, though, times were hard both economically and physically. Incomes were lower than the national average for manual workers, and labour-saving mechanisation, in the form of tractors, was only just getting underway. Veterinary help was expensive, so many problems were dealt with locally when livestock became ill. Remedies for sick cows apparently included bottles of stout, coffee and Epsom salts. Farmers lease or own 70 per cent of England's landscape, but the number of people working on it has declined substantially. In the 1930s, about

(a)

(b)

Figure 5.1. (a) Farming in the 1930s. *Source*: Mirrorpix/Getty Images. (b) Farming in the 1980s. *Source*: The Creative Drone/DigitalVision/ Getty Images

800,000 people toiled on the land, but by 2013 there were fewer than 200,000 (Zayed & Loft, 2019). Following the Second World War, government intervention in the industry increased substantially, and so did farming productivity, which more than doubled between 1950 and 2018. The Agriculture Act of 1947 was transformative; it guaranteed prices to farmers for whatever was grown and gave assurances that their land would not be taken away. Since then, the economics of farming in Britain have been hugely influenced by external factors, initially including guaranteed markets but then by a range of varying and unpredictable financial subsidies. This Fred Karno's circus has underpinned agricultural intensification with all its damaging implications for wildlife.

Farming Economics

Since the 1950s, the farming enterprise in Britain has relied hugely on taxpayer support, grossly distorting the relationship between the costs of food production and the prices paid by consumers. Price guarantees were made up by government as and when supply exceeded demand, as increasingly happened when the new technologies increased yields of crops and livestock. A further complication was that neighbouring countries also gave taxpayer support, at varying levels, to their own agricultural industries. Comparative costs of home-grown and foreign imports were therefore skewed by the ranges of subsidies elsewhere rather than just by local production efficiencies, leading to pressure for selective import controls to protect UK farmers. Entry to the European Economic Community in 1973, later the EU, essentially set a policy of farming subsidies in stone. The Common Agricultural Policy (CAP) became a mechanism for supporting agriculture across the community and achieved notoriety by distorting production to generate butter mountains, milk lakes and the like.

CAP funding comes in two forms. The single payment scheme (SPS, later the basic payment scheme, BPS) is by far the largest contribution to farming finances and is dedicated to maximising productivity with some associated environmental safeguards. The much smaller agri-environment payouts support operations specifically to protect wildlife. These latter are not classified by the government as subsidies because they compensate farmers for income lost by carrying out work benefitting the environment, such as planting woodland. The extent of support by SPS/BPS is staggering. Farmers make vastly more money from these subsidies than they do from farming itself. In 2015, on average English farms made a profit of £39,000 but only £2,100 of this came from agriculture. In the case of cereal farms, there were average losses of £9,500, discounting subsidies. The system only worked because the average English farm received £24,900 by way of government support in that year. These figures change annually and between various sizes of farm. Large holdings in 2014/15 made £22,300 from agriculture, whereas small farms lost £6,600. Sticking with averages, farms that produce mostly cereal lost £9,500 on their agricultural income but they also received a net £33,900 in subsidies. Broadly similar situations pertained in all the UK countries. The CAP underwent substantial reform in 2015, but the basic premises, with funds for food production and environmental protection (involving an expanded rural development programme), remained the same.

Agriculture in the UK is therefore underpinned by a bafflingly complex interaction between the value of home-grown farm products in competition with imports, and a varying array of taxpayer-based subsidies. This situation has created ongoing uncertainties and stress for farmers, with the profitability of what they grow changing unpredictably from year to year. Areas given over to wheat increased in the later years of the twentieth century, while barley declined. Since the mid-1970s, cattle numbers have dropped by more than 30 per cent, while sheep numbers have remained relatively stable after a short-lived increase in the 1990s. Arguments about the high quality of outdoor life enjoyed by farmers take little account of the intense workload, the highest rates of fatal accidents of any industry or that the average income for farmers in the UK is only around £10 per hour. The BPS has been administered in the UK with astonishing incompetence, leaving farmers waiting years for their entitlements. Some wildlife charities, one of which I have been associated with, are also eligible for these funds and have experienced similar levels of administrative ineptitude. So bad had this situation become that financial uncertainties have left some famers so desperate as to commit suicide.

5.3 Agricultural Intensification

Decades of Change

Picture a typical English lowland farm in the 1930s. Fields were small, typically up to a couple of hectares, bounded by hedges maybe 2 m thick. Many had corner ponds for watering sheep and cattle. Mixed farming was the norm, including livestock and arable, the latter often with a 3- or 4-year crop rotation. Fertiliser was mostly animal dung, and weeds and insect pests just had to be tolerated. Wildlife abounded. Thrushes and blackbirds nested in the hedges, the hayfields in summer were a blaze of colourful wildflowers, and rabbits and hares ran amok. Skylarks soared above, corncrakes scurried through the herbage and lapwings congregated in damp hollows. Amphibians and aquatic insects abounded in the ponds. I am just old enough to remember places like this, although by the time of my youth in the 1950s and 1960s, they were already rare. Almost all of these farm features were on the way out, and the British countryside would never be the same again.

The SPS/BPS has paid farmers per hectare of land under cultivatable condition, irrespective of what is grown on it. One consequence of this system has been increased exploitation of marginal land, which, without subsidies, would be uneconomic to farm. Those steep, scrubby slopes that previously provided refugia for wildlife in the farming landscape are no longer safe, and I have known such places being reclaimed almost as a matter of pride. However, this activity is small beer compared with the main, devastating procedures underpinning agricultural intensification, all of which are dedicated to increasing food production.

One by one the farmland features necessary to support a rich wildlife community have been degraded or lost. Increasingly sophisticated, and often huge, new machines such as combine harvesters came into play and rendered small fields an inconvenient nuisance. Arable fields became huge, covering tens of hectares or more, so that with fewer edges the harvesters waste less time turning round. The English prairie was born. A necessary condition of that change was removal of boundary hedges. In some areas, 50 per cent of hedgerows existing before the last war were grubbed out. Many thousands of kilometres of this important habitat were lost in the late twentieth century, and although policy has recently changed with some replanting, this has nothing like replaced what has gone. In addition, hedgerows today are often flailed to within a centimetre of their life, leaving them much thinner and less valuable to wildlife than their forebears. Ponds, like hedges, became a waste of space in arable farms and even in pastures, where piped water increasingly replaced them to quench the thirsts of livestock. Many were infilled deliberately or left to silt up by natural succession. Dew ponds on the South Downs, once famous as watering holes for sheep, were few and far between when I surveyed them in my part of Sussex 40 years ago. Over 70 per cent of those marked on ordnance maps were no more than damp or dry hollows. This was an all-too-typical situation; according to DEFRA, the UK lost at least 75 per cent of its ponds during the twentieth century and once again recent attempts to increase numbers, including the Million Ponds Project, have a long way to go in making up for depletions of this highly biodiverse habitat.

Of no less importance than physical changes on farmland were goings-on in the fields. Autumn planting, particularly of wheat and oilseed rape, became popular because growth starting in spring gave the best yields. Increasingly mild winters have accentuated this early growth spurt but also generated problems with competitive weeds such as black grass (*Alopecurus myosuroides*). Historically, stubble was commonly left over winter out on the fields and planting of the next crop delayed until spring. The change to autumn planting was bad news for some wildlife, especially finches and other birds that used stubble and insects lurking among it as a winter food source. Another undesirable practice took hold in many arable regions, notably the burning of stubble in late summer to kill off insect pests. Clouds of black smoke blossomed over the Sussex Downs where we lived during the peak of this malpractice in the 1980s, a truly dismal sight that eventually ceased under pressure from government and the non-farming community.

Dominating all of these developments have been the greatly increased applications of agrochemicals, including artificial fertilisers and pesticides, aimed at increasing crop yields well above historical norms. Realisation that plant growth is limited by nitrates and phosphates led to mass production of both chemicals for use on farms. More than 3 million tonnes of ammonium nitrate, produced on an industrial scale, are used every year in Britain, while rock phosphate, a natural resource, is treated chemically to produce calcium phosphate fertiliser. These materials are not innocuous. The potentially explosive nature of ammonium nitrate is a downside that has been exploited by terrorists and has also caused many inadvertent deaths. Accidental detonation in Beirut of nearly 3,000 tonnes of stored nitrate in 2020, which killed more than 200 people, was a dreadful example of the dangers posed by this chemical. From the wildlife perspective, the main issues from overuse of fertilisers have been damage to wild plant communities and the contamination of water courses. High soil nutrient content promotes the success of competitively superior plants, especially grasses, at the expense of many less vigorous (but beautiful) flowering species. Eutrophication of ponds from excess nutrients promotes algal blooms and decline or extirpation of the higher plants on which the ecosystem depends. Run-off into rivers has become one of the main sources of pollution in waterways in addition to sewage inputs. Again, the upshot is a reduction of both plant and animal diversity.

Last but not least, there are pesticides. Extra nutrients provided one string to the crop growers' bow, but control of competitors and predators became another. Herbicides to kill off arable weeds as well as fungicides and insecticides to wipe out organisms parasitising or eating plants destined for human consumption have all become part of the chemical cocktails distributed across vast areas of farmland in the UK. Following withdrawal of some previously popular herbicides, phenmedipham and desmedipham have become the backbone of weed control in Britain and elsewhere in Europe. They are typically used in low doses as mixtures with other herbicides to control a spectrum of weeds, including, of course, the UK's panoply of wildflowers. Initially pesticides were welcomed uncritically and used in houses as well as outdoors. I recall my parents distributing DDT powder, one of the earliest insecticides, freely on and under carpets in our home during the 1950s wherever some unlucky fly or spider might lurk. It took decades of research on this and on later generations of polychlorinated biphenyls (PCB) pesticides to reveal serious downsides of their unregulated applications. Many are highly persistent, degrading only very slowly, and they rapidly permeated food webs from invertebrates to birds and mammals, including humans. Being fat soluble, these compounds are stored in the body's fat cells and accumulate at ever higher levels up the food chains. After lengthy rearguard actions by the producers, PCB sales in western countries have been curtailed, but they are still widely available in developing nations. But this was not the end of the matter. Neonicotinoids, a recent and chemically novel type of insecticide, have been employed since the 1990s and yet again have caused serious ecological problems. Astonishingly after the experiences with PCBs, these compounds were allowed onto the market with minimal testing of side effects, although not without public protests (Figure 5.2). Pollinator populations have taken a downturn in many countries following applications of these chemicals, but it has taken years to start enforcing controls on neonicotinoid use. And once again, agrochemical companies have challenged bans in court in the face of damning research, some of which they funded, demonstrating the damage being done. In 2020, the largest neonicotinoid-producing companies had combined incomes exceeding £20 billion. Conflicts of interest between shareholders and environmental protection seem all too obvious. Campaigning against restrictions on their products long after environmental dangers were confirmed has ignominious parallels with the disgraceful responses of tobacco companies to revelations about their toxic outputs, which continued for decades.

Figure 5.2. Protesting against neonicotinoid pesticides. *Source*: Thomas Spender/Moment/Getty Images

It is not only arable land that has been subjected to agricultural enhancement. Grassland in the hills and lowlands of the UK, mostly pastures for livestock, has been 'semi-improved' by addition of fertilisers, or 'fully improved' by ploughing and reseeding with yet more fertiliser. In pastures planted with rye grass (including *Lolium* spp.), the result is a dull green sward, a monoculture devoid of anything remotely interesting. 'Improved' hay meadows no longer undergo a single summer harvest but sustain two or three cuts each year to generate silage as a food supply over winter for cattle and sheep. This dismal development has further bleached the colour out of our countryside.

Upland Farms

The semi-mountainous regions of Scotland, Wales and northern England have long been strongholds of livestock farming. In many ways, they have fared better, from the wildlife perspective, than the intensively managed lowlands. Upland grazing is essential to maintain the beautiful scenery that characterises these places, but the recent spectre of overgrazing by increased numbers of sheep posed a serious threat to upland habitats. In 2008, among those upland SSSIs in unfavourable condition,

31 per cent of upland heathlands and 58 per cent of blanket bog were suffering from overgrazing. Too many sheep, in particular, allow coarser, less palatable plants to encroach. Vegetation fails to recover and sensitive species are lost. Ground-nesting birds preferring high sward levels suffer from reduced availability of nesting sites and from increased vulnerability to trampling when high stock densities occur. Small mammals lose ground cover, thereby increasing their risk of predation. Overgrazing was largely a result of farmers responding to policy initiatives designed to increase stock numbers. Introduction of the Hill Livestock Compensatory Allowance in the 1970s introduced payments based on stock numbers, and further schemes for beef and sheep, also in the form of headage payments, were introduced in the 1980s. These provided farmers with a financial incentive to overgraze their lands. Fortunately, policy and associated subsidies changed and stocking levels have reduced, ironically towards a point where undergrazing may become problematic in some areas. Upland farming has certainly undergone some traumas in recent decades with unfortunate implications for wildlife, albeit less severe than what has happened down the hill.

How Well Has Intensification Worked?

The post-war agricultural revolution has had considerable success in achieving what it set out to do. Overall farm productivity increased steadily from the early 1950s, more than doubling by the turn of the millennium but changing little since then (Zayed & Loft, 2019). Wheat yield increased continuously from around 2.5 tonnes per ha in 1950 to 8 tonnes per ha in 2000. However, there have been no further consistent improvements in yield over the past 20 years. And there has been a price to pay for intensification, especially in terms of soil loss and quality. Autumn crop sowing leaves bare fields over winter, which, together with the impacts of heavy machinery, causes large quantities of soil to be washed away under heavy rain. We experienced a dramatic example of this in Sussex, when a downpour left nearby roads and much of a local village under a thick blanket of silt washed off adjacent fields. Losses of valuable topsoil have been considerable. In one west-country river catchment alone, these were estimated at more than 5 tonnes per ha in a single year. Peat depth in the East Anglian Fens, one of the most productive agricultural areas in the UK, has diminished dramatically over decades partly because improved drainage allowed the peat to dry out and blow away. The government has belatedly recognised the seriousness of soil damage, with Michael Gove warning in 2020 that 'the UK is 30 to 40 years away from the fundamental eradication of soil fertility in parts of the country'. It is increasingly clear that modern, intensive agriculture is not sustainable in the long term even in the context of food production, even without considering the damage it has caused to wildlife and the transformation of beautiful to boring countryside vistas.

5.4 Farming and Terrestrial Wildlife Declines

Three consecutive State of Nature reports (Burns et al., 2013; Hayhow et al., 2016, 2019) consistently highlighted unrelenting wildlife declines in the UK and indicated clearly that agricultural intensification has been the most important cause of them. The ramifications of these farmland changes have spread across virtually all the taxonomic groups of British wildlife, leaving precious few unaffected.

Flowering Plants

The flashes of summer colour once ubiquitous in fields across Britain have become a joy of the past, memorable to elderly generations but virtually unknown to those growing up since the 1960s. The answer to Pete Seeger's emotive lyrics 'Where have all the flowers gone?' is, in the UK, mostly under the plough. Between the 1930s and 1980s, 97 per cent of English and Welsh lowland semi-natural grasslands were converted to something more productive, and losses have continued since that time (Peterken, 2013). Accompanying physical destruction of grassy meadows has been the wholesale application of selective herbicides, of which a staggering array is now on the market. Pretty much any plants that might compete with a growing crop, including all the pretty ones, can be dispatched with a toxic spray. So much for those gorgeous wildflower meadows with their dazzling arrays of sheep's sorrel (*Rumex acetosella*), yellow rattle (*Rhinanthus minor*), cow parsley (*Anthriscus sylvestris*), hogweed (*Heracleum sphondylium*), meadowsweet (*Filipendula ulmaria*), common knapweed (*Centaurea nigra*), oxeye daisy (*Leucanthemum vulgare*), dandelion (*Taraxacum officinale*), poppy (*Papaver rhoeas*), cornflower (*Centaurea cyanus*), ragwort (*Senecio jacobaea*) and the rest (Figure 5.3). Comparisons between botanical surveys before 1960 and those from 1986–1990 revealed huge range contractions of many once common and widespread arable weeds: pheasant's eye (*Adonis annua*), down by 67 per cent; corncockle (*Agrostemma githago*), thorow-wax (*Bupleurum rotundifolium*) and small bur-parsley (*Caucalis platycarpos*) all by almost

Figure 5.3. Wildflower meadow, once but no longer a common sight on UK farms. *Source*: Ashley Cooper/The Image Bank/Getty Images (A black and white version of this figure will appear in some formats. For the colour version, please refer to the plate section.)

100 per cent; cornflower by 99 per cent; narrow-leaved hemp-nettle (*Galeopsis angustifolia*) by 92 per cent; corn cleavers (*Galium tricornutum*) by 97 per cent; mouse-tail (*Myosurus minimus*) by 73 per cent; corn buttercup (*Ranunculus arvensis*) and shepherd's needle (*Scandix pecten-veneris*) both by 95 per cent; and spreading hedge-parsley (*Torilis arvensis*) by 93 per cent (Wilson, 1992). All of Britain's 49 orchid species suffered range contractions in the late twentieth century, with those on calcareous grasslands and in woodlands suffering the most. Eutrophication and changing patterns of grazing were largely responsible for the changes on the grasslands, with burnt orchid (*Orchis ustulata*) one of the worst affected, declining by 79 per cent (Kull & Hutchings, 2006). Probably 20 per cent of Britain's native wild plants are in steady decline. These are truly shocking results but sadly of little surprise to those that have watched on as the colours have leached out of a diminished landscape.

Invertebrates

Plants grown from seeds treated with neonicotinoids are effectively protected from many insect predators, but their impact is more far reaching. In particular, bees collecting pollen from treated plants become disoriented and eventually entire hive populations are lost. The declines of insects and other invertebrates as farming industrialised took longer to come to public attention than did the disappearance of wildflowers. Some people noticed that car windscreens gradually required less effort to remove splattered moths after a night drive, but only recently have the huge changes in insect numbers been quantified. Declines have occurred across all insect groups and have been substantially greater in Britain than in mainland Europe (reviewed by Hopkins, 2019), and in southern compared with northern Britain. Two main causes have emerged: the losses of so many plants on which insects feed, and the continuing applications of new-generation insecticides.

Butterflies are the insects best known to most people, flittering flashes of colour on a summer day. Farms, however, are not now the best place to seek such an experience. Numbers of many well-known species have dropped substantially over recent decades and by almost 60 per cent between 2000 and 2009 alone. In 15 out of 17 cases, declines on farmland between 1985 and 2012 correlated with use of neonicotinoid insecticides. Essex skippers (*Thymelicus lineola*), small skippers (*Thymelicus sylvestris*), small tortoiseshells and wall browns were the most strongly negatively associated with neonicotinoid applications (Gilburn et al., 2015). Moths have experienced similar and very large declines that confirm the car windscreen experience, while suction trap catches of dipteran flies at Rothamsted Research centre, set in an agricultural area, fell by 37 per cent between 1974 and 2014. These insects constitute significant components of the food intake of insectivorous birds on farmland. Beetles, too, have had a hard time. Between 1994 and 2008, 75 per cent of carabids studied at 12 sites across the UK became less common, and in half of these cases the declines were 30 per cent or more. Pastures in western Britain were the worst affected whereas chalk downland actually showed beetle increases.

Pollinators, however, have become perhaps the most serious insect casualties of modern agriculture. These chemicals persist in pollen and have toxic and sublethal behavioural effects on insects feasting on the flowers. When honey bees (*Apis mellifera*) collect pollen from treated plants, they become disoriented, eventually leading to entire hive populations being lost. But there are many more victims, including solitary bees, bumble bees, hoverflies and other regular visitors to summer blooms. This has consequences not just for pollinator diversity but also for yields of crops that depend on them. Dave Goulson and his colleagues at the University of Sussex have spearheaded research demonstrating that neonicotinoids are the main villains behind many insect declines (Goulson & Nicholls, 2016). And it has become clear that their effects are not confined to insects. Goulson and others have also implicated neonicotinoids in farmland bird declines, yet the farming community remains reluctant to face so many unpalatable facts. In Goulson's words, 'The ongoing complacency of NFU [National Farmers Union], and their willingness to aggressively attack scientists who speak out against overuse of pesticides does them no credit. Farming bird populations are collapsing, along with most other farmland wildlife. Isn't it time the NFU faced up to this, and took some responsibility?' (Goulson, 2017). Studies in Germany revealed that insect declines in that country have spread beyond farmland and are unexpectedly severe in protected areas, possibly due to pesticide sprays blowing well beyond the application sites. There have long been complaints from householders in Britain about sprays on adjacent fields looming over their gardens, confirming a possible mechanism for the wide impacts of these toxins. This is a dispiriting tale that continues 60 years after Rachel Carson's warnings. Silent springs, and summers, still haunt farmland wildlife.

Insects have been relatively well studied on farmland but other invertebrates much less so. For one group, though, the arachnids, there is circumstantial evidence of decreasing abundance since intensification began. Surveys prior to 1945 by William Bristowe yielded estimates of around 5 million spiders per hectare in a Sussex field of rough grass. No recent direct comparisons with Bristowe's study are available, but multiple surveys on organic and non-organic farmland in the northern hemisphere revealed that spider species diversity and abundance was 15 per cent and 55 per cent higher, respectively, on organic farms relative to intensively managed ones (Stein-Bachinger et al., 2020). It seems likely that other data-free invertebrate groups have suffered similar fates.

Farmland Birds

By the late 1960s, many birds of prey had disappeared from much of Britain. Sighting a sparrowhawk (*Accipiter nisus*) became a notable event in much of the country, and for buzzards only the far west remained as a safe refuge. All of this stemmed from gung-ho PCB use. At the summits of food chains, raptor tissues accumulated damaging concentrations of pesticides. As a result, growth of their egg shells was impaired, leading to widespread nest failures and thence population crashes. This discovery by outstanding ecologist Derek Ratcliffe paved the way for more cautious applications of pesticides although, as noted above, this realisation has certainly not been cautious enough.

Another very noticeable change over the past few decades has been diminution of many farmland bird populations. Indeed, this group of animals has been among the most publicly highlighted victims of modern agriculture, largely because of a thriving and vigilant bird-watching community. The BTO assessed changes in farmland birds between 1977 and 1991 (Milton, 1994) and between 1970 and 2013 (Newton, 2018), most of which amounted to alarming declines: redpolls (*Acanthis flammea*) down by 87 and 86 per cent (respectively, from the two studies), tree sparrows (*Passer montanus*) by 81 and 90 per cent, bullfinches by 73 and 40 per cent, corn buntings (*Emberiza calandra*) by 62 and 90 per cent, skylarks by 53 and 60 per cent, reed buntings (*Emberiza schoeniclus*) by 46 and 32 per cent, linnets (*Linaria cannabina*) by 36 and 57 per cent, goldfinches by 33 per cent (but up 154 per cent, according to Newton, 2018), house sparrows by 23 and 66 per cent, and yellowhammers by 19 and 55 per cent. Trends for most of these and for other species have continued along similar lines over subsequent years, with the notable exception of goldfinches, which have thrived recently, probably since they caught on to garden feeders. Corncrakes (*Crex crex*), common in fields across the country a century ago, have become more or less confined to Hebridean islands. Turtle doves and nightingales are closing in on extinction in the UK. Summer migrants including swallows, swifts, house martins and cuckoos are much rarer heralds of spring than they were in my youth. Yellowhammers departed our Somerset village many years ago, almost certainly consequent on food shortages caused, as elsewhere, by pesticide applications (Hart et al., 2006). It is notable that the less intensively managed countryside of Scotland has suffered fewer and less severe declines of many farmland birds. Cuckoos still resonate on the hillsides, and house martins are much commoner there than in southern Britain. Walking the footpaths of an English lowland farm, by comparison, is less enjoyable for birdwatchers than was the case 50 years ago. Plenty of corvids and pigeons but far fewer skylarks reward today's countryside rambler.

Other Vertebrates on Farms

The disappearance of so many field ponds (Figure 5.4) coincided, unsurprisingly, with that of many amphibians. Some species, such as common frogs (*Rana temporaria*), common toads and smooth newts found new

Figure 5.4. Terminal decline of a farm pond. Photograph by Joe Costley/ Plantlife, UK (A black and white version of this figure will appear in some formats. For the colour version, please refer to the plate section.)

accommodation in garden ponds but not on a scale comparable with what had been lost. Great crested newts have been especially disadvantaged because garden ponds rarely meet the conditions they require for successful breeding. The disappearance of amphibians from so much of the countryside was recognised and quantified early on (Cooke, 1972). Reptiles, too, have had a bad time on intensively managed farmland. Viviparous lizards require undisturbed areas of cover and plentiful invertebrate food. Both are in short supply on most modern English farms. On the Sussex Downs where we lived for over 30 years, these lizards, abundant in the 1970s, diminished to nothing by the 2000s on and around the field and park opposite our house. Adders are in continuous decline and are now rarely seen on farmland. Other observers have had similar experiences, while populations on more typical reptile habitats such as heaths and scrubby common land have generally fared better. All this is anecdotal evidence, but the disappearance of relatively conspicuous animals in agricultural settings where they were recently common begs an explanation. The East Sussex Downs still offer pleasant distant views across the English Channel, but much of the landscape has become an arable desert lacking in reptiles. It was not always thus. Describing this same downland a century ago, Hudson (1900) recounts that 'the common lizard is found everywhere among the gorse and heath' and that 'the adder is common in suitable places, on the high slopes, especially where the gorse bushes grow mixed with heath and tussocky grass. . .a good many sheep die of adder bite'. Grass snakes (*Natrix helvetica*) are the reptiles most often seen on English farmland, especially in areas near water, and in this case there is no evidence for steep declines, although that may simply reflect a lack of evidence of any kind.

Among the mammals, hedgehogs are undoubtedly the species most intensively studied during their ongoing and precipitous recent declines. Despite this attention, explanations for the hedgehog's plight remain uncertain, but, conspicuous as they have been, road deaths are considered unlikely to be a major cause. The main reasons are thought to include the continuing intensification of agriculture, with reductions in permanent pasture and losses of hedgerows and field margins. We know even less about the impacts of intensive agriculture on other small mammals. Field and bank voles (*Microtus agrestis* and *Myodes glareolus*) as well as mice species are still generally abundant, although concerns have been raised anecdotally about falling numbers of voles in arable areas. Many bats have declined in the UK over the past 50 years, although some species have partially recovered recently as a result of conservation programmes. Bat declines have had several causes, one of which has certainly been a reduction in the numbers of the insects they prey on over farmland (www.bats.org.uk/about-bats/threats-to-bats). For most mammal species, though, we don't know how they have responded to farming changes. Only in the cases of foxes and badgers is there confidence that numbers are stable or increasing across most of the UK. Somewhat ironically, badgers have benefitted from increased plantations of maize, an environmentally damaging cash crop that this omnivore particularly relishes.

5.5 Impacts of Farming on Aquatic Habitats

The Quality of Freshwaters

In addition to continuing problems arising from sewage outflows, rivers across Britain have been extensively damaged by run-off and effluents from farms. Nitrates and phosphates enhance crop plant growth substantially and are the main constituents of fertilisers used by farmers since intensification programmes began. Unfortunately, these chemicals are not entirely taken up by the crops, and substantial quantities leach out into rivers and other waterways draining fields. Remaining ponds are also prone to contamination from these nutrients, which promote major ecological change by the process of eutrophication. Growth of generalist, vigorous competitors, especially algae, shade out higher plants, the loss of which in turn diminishes invertebrate numbers and diversity. Water becomes deoxygenated by night-time respiration of prolific algal growth to the point where fish and amphibians cannot survive. Damage is frequently enhanced by soil erosion, causing excessive quantities of silt to wash into rivers smothering weed beds and the gravels used by spawning fish. Efforts are underway to reduce farm-based pollution, but as recently as 2015, only 20 per cent of water bodies in England were classified as in good condition. All of this is yet another indictment of poor regulation that has serious consequences for wildlife (Hopkins, 2018).

Nowhere better exemplifies the impact of chemical pollution in freshwaters than the fate of the Norfolk Broads, a complex ecosystem of small lakes and interconnecting channels. As a student in the late 1960s, I boated on a crystal-clear Hickling Broad, looked down onto vigorous growths of higher plants and netted out fascinating invertebrates including mysid shrimps (*Neomysis integer*), inhabitants of freshwaters with a present or past saline influence. I was just in time, because steadily increasing eutrophication collapsed the aquatic ecosystem at Hickling, and many other Broads, during the 1970s. There has been limited recovery of water quality since then, but most of the Broads' waterways are now muddy channels with little to see beneath the surface. Happily, above the water they retain much of their historical wildlife interest, including swallowtail butterflies, bitterns and cranes.

Across the country, especially in lowland England, nitrate pollution will remain problematic for decades to come, and not just for wildlife. Water companies are wrestling with expensive methods to remove nitrates from drinking water, which at concentrations now present are a significant health hazard. Maybe this problem emerging for humans will prompt more substantial action to tackle a largely unseen threat to our precious environment. As well as nutrients leaching into our waterways, pesticide run-off including neonicotinoids is also a problem with unknown consequences for aquatic invertebrate and maybe for humans too.

Drainage and Abstraction

Infilling of ponds and pollution are not the only damaging activities wreaked by farmers on the country's freshwater systems. There is a long history of land drainage in Britain, starting with the Romans 2,000 years ago and continuing apace for centuries afterwards. The well-known efforts of engineers in the seventeenth century transformed the East Anglian Fens from a system of lakes and islands into the arable plain we see there today. Overall, the UK has lost perhaps 90 per cent of the wetlands present in pre-Roman times, but this has mostly involved low-lying landscapes easily converted for farming activities. Fen drainage continued to 'improve' after the 1950s, and it is probable that increased dredging and canalisation of ditches and rivers in that region lead to the demise of the burbot (*Lota lota*), the only freshwater fish to have become extinct in the UK within historical times. Starting in the last war, however, large areas of land not previously considered worth draining were transformed by new technologies using permeable pipes laid below the soil surface by specialised machinery designed for the job. The area of farmland dried out in this way rose from 15,000 ha in 1950 to more than 110,000 ha by 1975 (Robinson, 1990). Across much of Britain, fields prone to occasional winter flooding have become a thing of the past. Alongside drainage has been the increased protection of land adjacent to rivers from flooding, especially along tidal reaches, by the construction of embankments preventing occasional overflow. Inevitably, these changes have impacted on wetland wildlife. Flocks of lapwings, once common sights in an early spring walk wherever water lay on the ground, are increasingly consigned to protected areas where drainage is constrained. Between 1970 and 2018, a third of 26 species of wetland birds declined continuously and lapwings have been among the most seriously affected. One of the largest populations of natterjack toads formerly abounding along the River Yare floodplain in East Anglia was wiped out in the 1950s by embankments and arable conversions that abolished their shallow breeding pools.

When not draining land, farmers regularly need to abstract freshwater in summer to sustain crop growth. Neither too much water nor too little is the aim, a goldilocks zone aspired to by agriculturalists around the world. Water sources include rivers and bore holes dug down to underground aquifers, both of which can have serious consequences for wildlife if overused. Having spent my earliest years amidst the drizzle of South Lancashire, it came as a surprise to realise that even in our well-watered islands there can be times of drought. This is particularly true of south-east England, an area with some degree of rain shadow relative to the north and west but which is also home to large areas of thirsty arable land and a high density of equally needy humans. Overall water abstraction in England changed little between 2000 and 2017, at just over 10 billion m^3 per year. Most of this goes into domestic supplies or for industrial use, with only about 10 per cent of it taken by farming activities. The situation in Britain is therefore very different from the cereal-growing prairies of the USA and northern India, where bore holes required for crop irrigation have consistently lowered water tables towards, before too long, a point of no return. An increasing proportion of the UK's root and vegetable harvest, including 36 per cent of UK potato fields, requires irrigation, but it is domestic rather than agricultural use that is imperilling water supplies in parts of England. There have been adverse consequences for wildlife where water supplies have been overexploited. Waterbournes, the headwaters of streams fed by springs emerging from chalk hill aquifers, have dried up with increasing frequency. The rich plant and invertebrate communities of this high-quality water supply have been badly damaged, and the spawning grounds of salmonid fishes reduced in extent. The chalk-stream rivers of lowland England are an internationally rare and extraordinarily attractive habitat, with crystal-clear water, undulating growths of water crowfoot (*Ranunculus aquatilis*) and trout posturing effortlessly in the flow. Increased abstraction from the South Downs National Park caused the River Lavant to cease running altogether in 1994, 1996 and 2003, and other rivers have suffered a similar fate, with 25 per cent of them overexploited in 2017. Borehole usage in Suffolk's Redgrave and Lopham Fens came close to exterminating one of only three known populations of the fen raft spider (*Dolomedes plantarius*) in the 1990s, before the hole was closed (Smith, 2020). This problem cannot be laid primarily at the feet of modern agriculture, but water abstraction for whatever reason has affected wildlife negatively in the countryside.

Farming and the Marine Environment

Farming no longer stops at Land's End. One strategy devised to reduce pressures on wild fish stocks is the use of aquaculture to increase production of edible species in efficient, controlled conditions. This method, mostly using offshore containers, increased rapidly in scale from the 1970s onwards. Seascapes off the west coast of Scotland and associated islands have become dotted with floating cages holding tens of thousands of Atlantic salmon. This seems like a good idea, but the devil is in the detail and the practice has not proved ecologically sound. While dramatically increasing the supply and decreasing the costs of these tasty fish to consumers, there has been a serious price to pay by wild populations of the same species, and of migratory sea trout (*Salmo trutta*). Enormous numbers of parasitic sea lice escaping from farmed salmon cages have infected wild fish migrating past them with devastating consequences, and escapees from storm-damaged cages have interbred with their wild relatives, weakening their genetic constitution. The overall result has been a dramatic decline of wild salmon and sea trout along vast stretches of a coastline that once supported buoyant populations of these marvellous creatures (Tett et al., 2018). Too many Highland rivers and streams now tumble seaward devoid of one of the UK's dramatic wildlife spectacles, those waterfalls and weirs assaulted every year by leaping migratory fish en route to their spawning sites. Deep, clear but now empty pools and glides of pristine rivers festoon western Highland streams that once supported quivering shoals of these wonderful animals. The problems with this type of aquaculture extend well beyond the areas around the farms. The rapid expansion of offshore Atlantic salmon farms has generated a large market for sand eels and other small fish as food for the huge numbers of penned salmon. These small species have been harvested on an industrial scale to feed the big ones that humans eat, causing major disruptions of marine food webs. Sand eel fisheries are a case in point. Commercial catches, mainly by Danish boats, take almost half a million tonnes of sand eels every year, incidentally depriving seabirds such as puffins (*Fratercula arctica*) and kittiwakes (*Rissa tridactyla*) of essential food supplies for rearing their young. Exacerbated by negative effects of climate change on sand eel populations, the result has been serious prey shortages precipitating major declines of coastal seabird colonies that many people flock to see (Beebee, 2018). This ongoing disaster has continued unabated, largely ignored by politicians seemingly more interested in the value of fish exports and employment of the rather few people engaged in the industry.

As described in Chapter 3, depletion of marine fish stocks by human predation has caused serious declines of several species popular on UK dinner tables. However, predation is only part of the problem. Although not strictly a farming activity, collateral damage to the inshore seabed due to industrial-scale fishing is huge (Figure 5.5). Modern, high-tech fishing gear lays waste to large stretches of diverse marine habitats. Bottom trawling, in particular, is highly destructive (British Sea Fishing, 2016). Heavy nets are dragged across the sea bed and cause havoc on an appalling scale. Shellfish, crab populations, seagrass, weed beds and marine reefs are destroyed, and it is estimated that for every kilogram of marketable fish caught by this method, some 15 kg of peripheral marine life is killed. Large areas of many sea floor regions that have been bottom trawled repeatedly for years have been transformed from zones of high biodiversity to barren, featureless wastelands. Scallop dredging probably has the most severe ecological effect of all UK marine fisheries. The toothed dredges, weighing more than 2 tonnes, penetrate up to 10 cm into the sea bed often with repeat tows wreaking havoc in the target area. No habitat is safe, and damage extends over rocky reefs, sands, gravels and cobbles. Recovery takes anything from 1 to 10 years, assuming the site is not revisited, which it all too often is. Prawn trawling can also be a hugely wasteful process that has a significant impact on marine ecosystems through the amount of fish, shellfish and other marine species accidentally caught as bycatch. In the Clyde estuary, between 66 per cent and 80 per cent of the catch (by weight) taken by prawn trawling was unwanted and many of the discarded fish are edible species including hake (*Merluccius merluccius*), sole species, whiting (*Merlangius merlangus*) and haddock. The state of heavily trawled sea floors is truly shocking, with complex and attractive habitats converted to submarine deserts on a scale that would generate outrage if carried out on land. Imagine open-cast coal mining but on the scale of entire counties. This invisible cost of intensive fishing on British wildlife, largely unknown to coastal visitors enjoying grand views of apparently pristine seas, has gone on far too long.

5.6 Forestry and Wildlife in the UK

Woodland is surely a good habitat for wildlife, and in most circumstances this is certainly true. Comparable to wetlands, however, the UK has progressively lost the vast majority of the wildwood that spread across the land during the post-glacial warming. Clearance for agriculture starting

(a)

(b)

Figure 5.5. Damage done by bottom trawling. (A) Before trawling. (B) After trawling. *Source*: Colin Munro, Marine Bio-images/Colin Munro Photography (A black and white version of this figure will appear in some formats. For the colour version, please refer to the plate section.)

around 6,000 years ago resulting, by the beginning of the twentieth century, in only about 5 per cent of Britain maintaining woodland cover. The need for timber products, exacerbated by high demand in both world wars, led to the formation of the Forestry Commission in 1919 and the subsequent initiation of large plantation programmes. Since then, plantations have increased the total forest area to 'mediaeval levels' at around 13 per cent of the countryside. This still compares unfavourably with many European countries, where the average extent of woodland is 37 per cent. From the wildlife perspective, this new planting policy is where problems started. Most of the new woodland was planted with conifers, including Scots pine (*Pinus sylvestris*), which grows fast, at high densities and on a wide range of soil types. Scots pine is a native tree, one of the earliest post-glacial pioneer species, which became restricted mostly to upland regions in northern Britain as the climate warmed and other competitively superior trees replaced it further south. In Scotland, natural pine woodland is home to a wide range of fascinating and sometimes rare wildlife, including red squirrels, pine martens, Scottish crossbills (*Loxia scotica*, the UK's only endemic bird), capercaillies (*Tetrao urogallus*), crested tits (*Lophophanes cristatus*) and pine hoverflies (*Callicera rufa*). The RSPB reserve at Loch Garten, in the largest surviving remnant of the ancient Caledonian Forest, is where ospreys bred for the first time as they began to recolonise Britain following their extermination in the early twentieth century. Sadly, this bountiful woodland is much reduced from its former extent, although planting efforts are currently underway to expand it by reconnection with isolated forest fragments surviving around the Cairngorms. However, the Forestry Commission has also indulged in the widespread planting of non-native conifers with commercially desirable properties, including various larches, Corsican pine (*Pinus nigra*), lodgepole pine (*Pinus contorta*), Norway spruce (*Picea abies*), Sitka spruce (*Picea sitchensis*) and Douglas fir (*Pseudotsuga menziesii*). These plantations are invariably less suitable for British wildlife than the country's native woodlands.

Pine forests can therefore, under the right circumstances, be important wildlife havens. Unfortunately, restoring wildlife habitat was not of concern to the Forestry Commission when their activities got underway. More important then was identifying areas where conifers would grow well in areas that for millennia had been without them. The heathlands of southern and eastern England proved ideal, and vast areas were planted up with Scots pine in Dorset, the Weald, Norfolk and Suffolk. As discussed in Chapter 4, heathland is prime habitat for a number of rare species largely dependent on it. Since the nineteenth century, some 86 per cent of England's heaths have disappeared due to urbanisation, agricultural encroachment and forestry plantation, the latter accounting for 45 per cent of losses since 1920. Among many examples, at 25 km^2, Bloxworth was one of Dorset's largest areas of heathland in the nineteenth century. Now most of it is under Wareham Forest (Figure 5.6), with tracts of conifers interrupted, happily, with heathery rides and small blocks of heathland where sand lizards and smooth snakes still survive. However, problems with conifers are not restricted to the areas formally planted because Scots pine readily seeds into adjacent heathland, spreading rapidly unless continuously checked by aggressive management. Volunteers, including me, have spent innumerable hours pulling up pine seedlings on heathlands in Cnut-like attempts to stem an inexorable invasive tide.

In Norfolk and Suffolk, the Brecklands were once vast areas of wind-blown sandy heath, again with tranches of specialised rarities including stone curlews, woodlarks (*Lullula arborea*), Spanish catchfly (*Silene otitis*) and spring sedge (*Carex caryophyllea*) among many others. The 45,000 ha of Breckland habitats are registered as SPAs and SACs, but 85 per cent of the heathland there disappeared between 1930 and 1980. Many of the

Figure 5.6. Wareham Forest, once an extensive open heath. *Source*: Simon Byrne/RooM/Getty Images (A black and white version of this figure will appear in some formats. For the colour version, please refer to the plate section.)

losses were due to the plantation of what became Thetford Forest, which at 19,000 ha is the largest lowland pine forest in Britain. Unsurprisingly, there have been extinctions in the Brecks including three species of orchid, starry breck lichen (*Buellia asterella*), pasque flower (*Pulsatilla vulgaris*) and spring cinquefoil (*Potentilla tabernaemontani*). At least some East Anglians maintain a sense of black humour about the fate of Thetford Forest, as when the *Suffolk Gazette* reported: 'The controversial rewilding project in Thetford Forest has been scrapped after a Norfolk family of four were eaten by bears.' The explanatory subheading 'You couldn't make it up' was presumably helpful to some readers.

Problems with pine plantations have not been restricted to lowland heaths. Naturally acidic upland soils have become even more so following conifer forest development in parts of Wales, Scotland and northern England. This happens because the trees scavenge acidic atmospheric gases and ultimately release them into groundwater. In Wales, more than 12,000 km of streams and rivers have been acidified, harming aquatic insects, fish and birds. Fortunately, declining emissions of pollutants such as SO_2 coupled with addition of lime in selected areas have slowly been reflected by reduced acidification (Ormerod & Durance, 2009), but recovery of these aquatic ecosystems remains incomplete. Among other remedies to acidification is the increasing replacement of conifer stands with other native trees that do not have comparable acidifying effects.

On top of all this, there has been a major conifer-based scandal. In a policy designed to generate tax breaks for rich investors, a scheme in the late 1980s promoted conifer plantation in northern Scotland's Flow Country. What this vast area lacks in dramatic scenery is more than compensated by the wide range of plants and animals that thrive there. Among uncommon birds to be seen on a visit to this part of northern Scotland are red-throated divers (*Gavia stellata*), black-throated divers (*Gavia arctica*), golden plovers (*Pluvialis apricaria*), golden eagles and merlins (*Falco columbarius*). There are places in Scotland that would benefit from more trees, but the Flow Country is not one of them. Bogs were drained by extensive ditching, and non-native conifers, unsuited to this base-poor wetland, were planted in huge numbers. Attempts by the Nature Conservancy to arrest this damaging afforestation were ultimately successful but at a cost to the organisation, which was broken up into separate, country-based conservation bodies by a less-than-delighted UK environment secretary. Efforts by the RSPB, including establishment of the Forsinard Flows Nature Reserve, have resulted in extensive rewetting of bogs by blocking drainage ditches and the removal of the conifers from more than 2,500 ha of Flow Country habitat. The assault on this unique landscape was a pinnacle of the damaging plantation policy, since which time the emphasis of forestry development has changed for the better. This came too late, unfortunately, for many previously interesting heaths and bogs still lying forlorn amidst close-packed conifers and sterile layers of pine needles.

The post-war increase in woodland cover, primarily due to pine plantation, has ironic undertones for the more ecologically valuable native tree landscapes that have continued to decline over the same period. Ancient woodland (defined as being at least several hundred years old) now amounts to no more than 2.5 per cent of Britain's land surface. This is the most biodiverse habitat type in the UK, but more is lost year on year. Woodland birds have declined over recent decades in a similar but less severe fashion than farmland species, although the causes for decreases in woodland birds are less well understood. They probably include overall losses of mature tree copses and hedgerows that some of these species also use, as well as changes in woodland management, The causes of ancient woodland decline include direct development and expansion of intensive agriculture, but the value of these historical features of the UK countryside are increasingly appreciated by people from all walks of life. It may be some time before politicians come to realise the impossibility of replacing ancient woodland overnight, but recent schemes for planting large numbers of native trees are certainly welcome. This new enthusiasm is driven largely by climate change concerns, however, and it will be important to ensure that new battalions of 'carbon sinks' are only created in places where other important wildlife habitats are not affected.

5.7 The Myth of Self-sufficiency

Food Supply

Any hope that the UK could become self-sufficient in food production faltered within 40 years of the 1947 Agriculture Act. In fairness, complete self-sufficiency was never expected, not least because some foodstuffs cannot be grown in the UK. Assessed as the proportion of food consumed that is produced by British farmers, the closest the country came to achieving self-sufficiency, at around 78 per cent, came in the early 1980s. This compared with less than 50 per cent in 1950. Industrialisation of farming therefore made a substantial difference, with continuous increases for three decades as agricultural intensification spread across the country. Food price inflation in the UK has matched but not exceeded that of the overall consumer price index (Yurday, 2021) and on average, in 2018, people were spending just 10.5 per cent of their relatively higher incomes on food compared with 33 per cent in 1950. However, since a peak in 1984, the proportion of self-sufficiency with respect to foodstuffs has fallen considerably, and in 2019 it stood at 64 per cent. In another way of looking at this trend, in 1984 there was enough food produced in Britain to feed the nation for 306 days. In 2020, that time span had dropped to 233 days. In that year, the UK was just 18 per cent self-sufficient in fruit, 55 per cent in vegetables and 71 per cent in potatoes, although the percentages of the last two of these commodities have fallen by 16 per cent since the turn of the millennium. Pressures on the UK's food security, not yet apparent due to buoyant imports, are nevertheless real and are here to stay. The increasing global population and changing consumption patterns are increasing demand for foodstuffs everywhere, a situation Britain cannot perpetually avoid (www.foodsecurity.ac.uk/challenge/uk-threat/).

There are ways in which self-sufficiency rates could increase beyond what is currently achievable. One, much vaunted in recent years, is to reduce meat consumption or, in a vegan world, abolish it and all animal products altogether. Land presently employed to grow fodder for farm animals could be given over to produce food for humans. Under this scenario, even more of the countryside would be turned into arable fields, possibly leaving some previous pasture to go wild (as discussed in Chapter 10). There are at least two problems with this solution. One is the fact that most people like to eat meat and are unlikely to concur with its total disappearance from their diet. Although a third of British people claim to have reduced or ended their meat consumption, production of it has increased continuously on all the world's continents, mostly due to increases in the human population. Per-capita meat eating in the UK increased steadily after the 1960s, slowing down a little recently but still much higher than it was 60 years ago. Although superficially attractive, left unmanaged the ecology of old pastures would change at the expense of many species that could not persist under rank grass, scrub or woodland. Conversion of pasture to arable land would certainly be more bad news for wildlife. Unimproved, stock-grazed pastures maintain much higher levels of floral and faunal biodiversity than do modern arable fields. Semi-natural grasslands are typically rich in flowering plants, often with more than 15 but sometimes up 40 species per m^2. They also provide breeding habitats for 48 species of butterflies, approximately 85 per cent of the British total, and for a wide diversity of moths (Office for National Statistics, 2018). Historically, arable farming also supported a wide range of plants and animals, and unimproved or minimally altered fields still do. Unfortunately, intensively farmed agricultural land is a wildlife desert, with only a few robust species managing to survive on it. To maintain and enhance the prospects for wildlife in Britain, it will be vital to maintain pastures and not to further extend intensively farmed arable land.

A second approach to achieving self-sufficiency in foodstuffs could be to drastically reduce wastage. Food prices in the UK, relative to income, are among the lowest in the world. An unfortunate consequence of this comfortable situation is that huge amounts of what people buy get discarded rather than eaten, a situation utterly unthinkable in the economic climate of wartime Britain that precipitated the agricultural revolution. Total wastage in the UK, from farm to table, amounted to an astonishing 9.5 billion tonnes in 2018. Household consumption accounted for 70 per cent of this squandering, and most of the discards were still perfectly edible. Recent years have seen a small decline in food wastage in Britain, but as long as prices remain low, the issue seems likely to remain serious. In practice, the combination of cheap food and widespread indifference makes reducing waste on a sufficient scale to affect self-sufficiency in the UK a distant dream, although one towards which we should certainly aspire.

Timber

British industry uses about 50 million tonnes of timber every year. Most of this is softwood derived from pine plantations, but altogether native sources produce only 10–20 per cent of this timber requirement. Between the 1920s and 1990s, UK conifer woodland increased 3.6-fold in area, from 328,000 to 1,521,000 ha, but this was nothing like enough to achieve self-sufficiency. Softwood imports, which vastly exceed home production, come primarily from Baltic countries. These are likely to remain the main sources to satisfy increasing demand for wood in Britain indefinitely into the future, as it will never be possible to produce sufficient home-grown softwood. Most timber is used in the construction industry, a seemingly limitless market in an era of rapid housing developments. This demand means that efforts to reduce the area of pine plantations in the UK in favour of conservation-friendly habitats will be constrained, running contrary as they do to the country's needs.

5.8 Overview

There is no doubt that agricultural intensification and, to a lesser extent, forestry policies have been far and away the main reasons why British wildlife is in its current perilous state. No taxonomic group of plants or animals has remained untouched by the havoc wreaked on our countryside by the industrialisation imposed on it in recent decades. To what extent have these changes been driven by increases in the human population of the British Isles?

Intensification of farming practices was set in train by concerns, precipitated by food shortages during the Second World War, that the UK could not produce enough food of its own to support a human population of about 50 million people in the early 1950s. By 2020, this population had risen to almost 68 million, without suffering any significant food shortages along the way. Although increased farm production contributed significantly to this achievement, for a while approaching overall self-sufficiency in the 1980s, that Holy Grail was never approached and the prospects for achieving it receded in the subsequent decades. The demand for food imports to meet the country's needs, never eliminated, has continued to be substantial. This reliance on imports is a cause for concern even without threats from circling U-boats. Global wheat production, after a long period of continuous increase, has been virtually unchanged since 2013. Overabstraction of underground aquifers in North America and India to provide irrigation for these major wheat belts means that even mere stability of global wheat production may soon be a thing of the past. Wheat is just a single example of a major foodstuff, albeit a very important one, but is an indicator of what is likely to happen with others. There are claims that world agriculture already has the ways and means to feed perhaps 10 billion people, but even if this promise is fulfilled, its implementation would cause even more environmental damage than was generated by the Green Revolution of the 1970s.

About 70 per cent of UK land is potentially available for agriculture, more than the average of just over 60 per cent in all countries around the world. Importing food at the level Britain currently does, to support a human population the UK could not otherwise sustain, is arguably both immoral and selfish. It is only possible because the global economy permits population growth in rich, developed countries at the expense of feeding those more deserving of extra sustenance. This kind of exploitation also occurs within countries, not only between rich and poor ones. One of the world's richest nations perpetuates such disparities of incomes that many of its citizens rely on free handouts from food banks to avoid hunger. That country is the UK. Long-term prosperity, national or global, surely requires these imbalances to change, but this depends on a level of political commitment, including population policies, thus far almost entirely lacking.

As for forestry, even here the damaging plantation policies of post-war Britain relate to human population size, although the connections are less obvious than those with food supply. The earliest twentieth-century concerns about timber were precipitated by demands for military materials following shortages during both world wars. Since those difficult times, timber use in the UK has been increasingly monopolised by the building trade. This has been an important prerequisite for the expansion of urbanisation, principally via house building, that has continued unabated for decades past, as described in Chapter 4. And this in turn has stemmed from an increasing human population's requirement for decent accommodation. More people need more wood and more trees, usually conifers, to keep the expansion going.

The relationship between human numbers and wildlife depletion in Britain is evident in the countryside as much as it is in suburbia. There are ways of mitigating damage already done by changes in farming and forestry practices, some of which are already underway and are discussed in Chapter 10. Thus far, however, they have not halted, yet alone reversed, the downward trend in UK wildlife diversity and abundance.

6 · *Climate Change, Disease and Disturbance*

6.1 Other Outstanding Problems

Urbanisation and agricultural intensification have been the major instigators of wildlife declines in Britain. As well as these and pressures from persecution, introductions of alien plants and animals have, in some cases, precipitated habitat changes and diminutions of native species. This problem, which started on a large scale in the nineteenth century, was alluded to in Chapters 2 and 3. It is an ongoing issue, and there are now laws in the UK that attempt to prevent further arrivals of incomers, although, inevitably, undesirable immigrants still get through from time to time.

Then there are pressures on British wildlife that have become significant relatively recently. Climate change is now influencing plants and animals in the UK, with ambivalent consequences. Some species are benefitting from warmer weather and new ones are arriving, but others are faring badly and a few may die out if current trends continue. This concern emerged, with increasing vigour, in the closing decades of the twentieth century.

Wildlife diseases are part of life all around the world, contributing under most circumstances to long-standing evolutionary trade-offs between hosts and pathogens or parasites. Cycles of disease or parasite outbreaks and host recovery are normal, typically with no long-standing consequences for either organism. The holly blue butterfly (*Celastrina argiolus*), a common and delightful insect of warm spring days, undergoes cyclical population booms and busts inversely related to numbers of the ichneumon wasp (*Listrodromus nycthemerus*). This parasite inserts its eggs into the butterfly caterpillar, upon which the wasp larvae feed before emerging as adults from a deceased pupa. What has changed in recent decades has been the accidental import into the UK of diseases from far and wide across the globe, and for which native species have little or no defence. Some of these new pathogens have had and are having major impacts on susceptible plants and animals.

And then there is disturbance, with ever more people out in the countryside walking, boating and even flying close to animals stressed into moving away from their previously quiet and secure haunts. This is an issue in need of more study, but there is already evidence that some species have suffered from disruptions of their normal behaviour patterns. Any countryside walk inevitably causes some degree of disturbance to nearby creatures, but there is no doubt that the scale of this intrusion has escalated as the human population has grown and become increasingly mobile. As with previous chapters, the effects of these recent challenges to British wildlife are considered in the context of people numbers and their role, if any, in exacerbating the impacts of these new factors.

6.2 Climate Change and Wildlife

General Aspects of Climate Change

Consequences of planetary warming have become the environmental issue of our time. Hardly a week goes by without news of one disaster or another linked to increasingly dramatic climate-based incidents. Storminess in particular, often linked to devastating floods, is reported from countries as far apart as India and the UK. Ever more powerful cyclones drench and damage the Philippines and southern USA. Huge wildfires devastate the Amazon basin, California and Australia, while droughts intensify in parts of Africa. Arctic ice is disappearing in summer, glaciers are retreating in mountain ranges all over the world, and ice sheets in Greenland and Antarctica diminish year on year. Inevitably, most of the concern about climate change focuses on its impact on humans. Cyclones and wildfires kill people and wreck homes. Sea-level rise increases coastal erosion and the loss of homesteads, and threatens entire communities on low-lying oceanic islands. Unlike the biggest environmental issues of the twentieth century such as pesticide abuse, acid rain and damage to the polar ozone layers, climate change is truly global and nowhere is safe from it. The Paris Agreement on climate change of 2015 set a political ambition to curb greenhouse-gas emissions, the primary cause of global warming, but this is a daunting task, and at the time of writing, atmospheric CO_2 concentrations are still rising.

Although the future safety of humans has been central to concerns about global warming, impacts on wildlife have also registered. The reduction in Arctic summer ice poses long-term threats to several species including polar bears and walruses, which rely on ice floes for efficient hunting or for haul-out resting. The toll of wildfires on Australian wildlife has been appalling, with more than 3 billion animals killed during the 2019–2020 outbreaks alone. Some rare species may have been exterminated and others, including koalas (*Phascolarctos cinereus*) have suffered severe local declines. It seems certain that there is more of this to come. Long-term effects of climate change on wildlife are hard to predict but may not be entirely gloomy. One of the vanguard researchers on the subject has suggested that global species diversity may actually increase as a result of it (Thomas, 2018), but other analyses are pessimistic and suggest substantial losses of both species and habitats (Nunez et al., 2019). A lot depends on how successfully polluting emissions are reduced, so we will have to wait and see.

Climate Change and British Wildlife

Plants and animals in the UK have responded to a warming climate, albeit less dramatically than those in some other parts of the world. Naturalists have noticed many changes over the years, but it is the long history of recording both climate data and wildlife in Britain that has allowed serious study of what has been going on. Temperature and rainfall measurements in the UK are available for many decades past, and the same goes for the distribution and behaviour of a wide range of plants and animals. From eighteenth-century countrymen such as Robert Marsham to twenty-first-century scientists like Tim Sparks and Chris Thomas, meticulous observations and rigorous scientific studies have provided an impressive overview of how our countryside has responded to a warming world. In this respect, Britain is better placed than perhaps any other country to assess the impacts of climate change on wildlife, which have involved virtually every taxonomic group for which data are available (Beebee, 2018). Two major consequences of temperature increases have been identified, notably altered phenologies, and changes in distribution and abundance.

Phenology

Phenology, the timing of life-cycle events, has seen advances in the start of spring activities including first flowerings, emergences from hibernation and breeding events more or less across the taxonomic board. Winter temperatures have risen more sharply than those in summer, precipitating changes for many species in the UK at the start of the year.

There is an online database, naturescalendar (https://naturescalendar.woodlandtrust.org.uk/), holding tens of thousands of phenology records, which are added to every year and constitute a hugely valuable data set for those investigating climate-related changes. Some timing advances are easier to spot than others. Within the confines of a garden, snowdrops and crocuses (*Crocus vernus*), among the first heralds of spring, are now regularly on display soon after midwinter. Their responses are typical of multiple plant species that, between 1984 and 2008, bloomed on average, between 2 and 12 days earlier than in any other 25-year period since 1760. Butterflies are another conspicuous type of wildlife and many of them, too, are on the wing long before what was normal 50 years ago. Brimstones (*Gonepteryx rhamni*) and peacocks (*Nymphalis io*), which overwinter as adults, now regularly flutter about in February or even January in some parts of the country. All 31 species with records from southern England were flying earlier in the 1990s than in the 1970s, some by as much as 2 months. Typical 'summer' invertebrates have also changed their ways, although mostly to a lesser extent than spring emergents. These include some butterflies but also several species of dragonflies and damselflies that now appear a week or two earlier compared with historical norms.

Many vertebrates are also doing things differently in the twenty-first century. The arrival of common frog spawn is another easy-to-see indicator that has responded to climate change, although not to the same degree as our less conspicuous newt fauna. Traditionally spring breeders, migrating to ponds in February or March, I watched dramatic changes in newt arrival times in our Sussex garden ponds. Between the late 1970s and mid-1990s, the first individuals of all three native species appeared ever earlier and eventually could be spotted before Christmas (Beebee, 1995). Britain's hugely popular bird-watching hobby has produced more phenological information on these animals than has accrued on any other animal group, providing the BTO with sufficient data to generate impressive analyses of climate effects. Most of the 36 species with sufficient records to assess started to lay eggs progressively earlier between the 1940s and 1990s. In the case of robins, there was a shift of 15 days between these decades. Arrivals of summer migrants such as swallows and house martins were also reported in advance of historically typical timings, despite the expectation that their journeys must be influenced by conditions during passage, and maybe also by weather on their overwintering grounds. Notwithstanding these potentially confounding factors, the changes were often substantial. The mean arrival times of pied flycatchers (*Ficedula hypoleuca*), for example, were typically around mid-May in 1974, but these birds were here by the start of that month in 2001. Reptile hibernation, at least for some species, is less intense than it used to be and adders have been photographed basking in every month of the year. For mammals, generally harder to observe routinely than amphibians and birds, there is more limited information, although hedgehogs and bats seem to be more intermittently active in winter, and probably emerge from hibernation earlier, than was normal in times past.

In most cases where phenological change has been documented, there are clear correlations with temperature, usually in the 1 or 2 months before the event. This strongly suggests that the advances so widely documented are indeed related to warming trends, particularly in winter. An important question that follows from these observations is whether altered phenology is having impacts on populations of the species concerned. Some consequences related to climate change have certainly been noted. In a study of 232 plants in Guernsey, mostly garden flowers, earlier blooming was associated with shorter flowering times. This might allow stronger somatic growth later in the year, or reduced pollen output, potentially with diametrically opposed effects on fitness. It could also be bad news for pollinators. Warm winters are having adverse effects on some birds, as judged by poorer body condition. Three species of tits – blue (*Cyanistes caeruleus*; Figure 6.1), coal (*Periparus ater*) and great (*Parus major*) – experienced long-term nationwide decreases in body mass between 1965 and 2017, correlating inversely with mean winter temperatures and with sparrowhawk abundance (Furness & Robinson, 2019). However, at least in the case of blue tits, there is no evidence of population declines; on the contrary, it is one of relatively few passerines still faring well in the UK, probably assisted by garden feeding stations (Stenning, 2018).

One anticipated problem with phenological changes has been the possibility of asynchronous responses among component species of food webs. A system that has attracted close scrutiny involves oak trees, moths, great tits and possibly sparrowhawks. Oak leaf bud burst in spring is followed by attack from caterpillars, especially those of winter moths (*Operophtera brumata*), which can become extraordinarily abundant. This event is followed in turn by mass predation of the caterpillars, huge numbers of which are needed to sustain the nestlings of great (and other) tits. These insects form the bulk of tit food in oak woods, the prime habitat for these birds. Somewhat later, newly fledged and inexperienced

Figure 6.1. Bluetits have suffered reduced body condition as a result of mild winters. *Source*: Gary Chalker/Moment/Getty Images

tits fall relatively easy prey to sparrowhawks. Timings are critical. If peaks of caterpillar infestation come too early or too late, mortality of tit chicks from starvation can ensue. The trees also benefit from synchrony, as heavy caterpillar infestations unregulated by tit predation can reduce fitness and reproductive output, meaning fewer acorns later in the year. Evidently, bud burst and egg-laying by moths and by great tits have all become earlier in response to climate change. In the UK, synchrony between these events has largely been maintained, with all advancing at similar rates. This has not, however, been true everywhere, and in the Netherlands, great tits failed to advance their egg-laying in concert with caterpillar abundance, seemingly because patterns of spring warming have differed between the two countries. Great tit breeding success in the Netherlands has declined somewhat, but populations have remained robust because their sizes are determined mostly by bird density and variations and by the autumn beech crop later in the year rather than by survival to fledging. Sparrowhawks have shown no particular response in their breeding times and are doing reasonably well, possibly because they have a broad prey spectrum not strongly related to any particular species. Thus far, there has been little evidence of phenological asymmetries causing wildlife declines, although as described in Chapter 4, the impacts of newts on frogs in garden ponds do seem to have increased as a result of asynchronous changes in breeding times. Because garden ponds are now such an important habitat for frogs, this increased vulnerability to newt predation could be having effects on their populations at a national level.

Distribution and Abundance – The Good News

A profound result of warming conditions in the UK has been substantial changes in the abundance and distribution of numerous plants and animals. As shown by Burns et al. (2018), these changes have more often been beneficial rather than damaging to British wildlife, such that the evolving climate has contributed, thus far, to increased biodiversity and more robust populations of warmth-loving species, particularly in England. But this is a fine balance that could eventually tip the other way, because there is also a suite of plants and animals that are adapted to cool conditions and that are already suffering from elevated temperatures.

So at a time when there is so much bad news for wildlife in Britain, it is reassuring to note positive consequences of climate change that have allowed some plants and animals, albeit a minority, to improve their lot. While most orchids are declining due to habitat destruction, some are going against the grain, increasing both their ranges and their population sizes. These include the rare lizard orchid (*Himantoglossum hircinum*) and the lady orchid (*Orchis purpurea*), both of which have increased their population numbers twofold or so, from very low bases, since the 1980s. Other plant beneficiaries, some of which seem to have broadened their habitat niches, include great lettuce (*Lactuca virosa*), green-flowered helleborine (*Epipactis phyllanthes*) and spotted medick (*Medicago arabica*). Invertebrates, too, have responded to rising temperatures, sometimes dramatically. Several butterflies including Adonis blues (*Polyommatus bellargus*), speckled woods (*Pararge aegeria*), commas (*Nymphalis c-album*), and holly blues have spread into new habitats and extended their distributions northwards. Much the same applies to moths. Pretty Jersey tigers (*Eupalagia quadripunctaria*), until recently a species of south-western coasts, is now widespread across southern England and turns up in our Somerset village every summer. Dragonflies and damselflies, dazzling signatories of sunny days, have been among the most astonishing respondents to increasing temperatures. Several species previously known mostly from southern England have moved north, all but three of the UK species extending their ranges in that direction by an average of 74 km since 1970. The impressively large emperor dragonfly (*Anax imperator*) is now a novel Scottish resident, along with several other newcomers to that country. And there are several recent arrivals from across the English Channel, not just visiting but breeding and surely now here to stay. Small red-eyed damselflies (*Erythromma viridulum*), first seen in Essex in 1999, are one such species, which since then have become widespread across much of southern and eastern England. Yet another invertebrate group with a few members responding positively to climate change is the Orthoptera. Long-winged and short-winged coneheads (*Conocephalus discolor* and *C. dorsalis,* respectively) have, among other crickets and grasshoppers, headed north in spectacular fashion since 1980. No doubt members of other less well-documented invertebrates have responded similarly with northward range extensions and colonisations from continental Europe, but it is important to bear in mind that most members of all these animal groups are in continuous decline, not enjoying rescues from warmer weather.

Some vertebrates are also benefitting from climate change. Once again responses are best documented for birds, and as with many invertebrates, the main effects of warming weather have been northward shifts in range distributions and the arrivals of new species. Many insectivorous birds head south for winter, but among those that remain in Britain, the Dartford warbler (*Sylvia undata*) kept, until recently, just a precarious toehold on the Dorset heaths. In the severe winter of 1962/63, it almost became extinct, but since then this charming but elusive warbler has thrived and extended its range as far as East Anglia. Once again, the BTO has provided the best evidence of climate impacts on birds, with 59 southerly species shifting their range edges northwards by an average of 20 km between the late 1960s and late 1980s. This trend continued at least until 2011 and even included increasingly rare birds such as turtle doves. Once again, there were convincing correlations with temperatures, in this case mostly summer ones. And new arrivals on British shores have been among the most obvious, and celebrated, results of climate change. On the Somerset Levels, I have been among many observers delighting in the appearances first of little egrets (*Egretta garzetta*), then of great white egrets (*Ardea alba*) and most recently of substantial flocks of cattle egrets (*Bubulcus ibis*), all since the 1990s. It is now unusual to visit the Avalon Marshes without seeing at least one of these immigrants, all of which have settled in as regular breeders. These are perhaps the best documented of Britain's new resident birds, but numerous others are visiting and in some case also breeding successfully. Other vertebrate responses are much more difficult to detect, but there is some suggestion that badgers have benefitted from milder winters and that the bat Nathusius' pipistrelle (*Pipistrellus nathusii*), once an occasional visitor to the UK, has established breeding colonies around the country.

There is no doubt that climate warming has increased Britain's biodiversity by facilitating invasions of species not previously resident, and by increasing the population sizes and distributions of some long-standing natives. The news, however, is by no means all so positive.

Distribution and Abundance – The Bad News

Increasing temperature is not universally beneficial to British wildlife. Within our shores are plants and animals adapted to cool conditions, particularly in the uplands of Wales and northern Britain. Many of these are feeling the heat and beginning to succumb in an increasingly unfavourable environment. Temperature increases between 1993 and 2007 were around twice as great at high altitudes compared with the lowlands. Specialised Arctic–alpine plants in the Scottish and Welsh highlands are at risk, including the spectacularly beautiful blue heath (*Phyllodoce caerulea*), alpine gentian (*Gentiana nivalis*) and Snowdon lily (*Gagea serotina*) (Figure 6.2). Surveys comparing abundances of these and other upland plants before 1970 with more recent investigations in the early 2000s identified many declines in the ground cover of rock sedge (*Carex saxatilis*), red-stemmed feather moss (*Pleurozium schreberi*) and snow-bed willow (*Salix herbacea*), among plenty of others. By contrast, some species, mostly generalists such as mat-grass (*Nardus stricta*) that are comfortable with warmer conditions, had moved uphill.

A similar story has unfolded for cold-adapted invertebrates. Among the upland butterflies, the northern brown argus (*Aricia artaxerxes*), Scotch argus (*Erebia aethiops*) and large heath (*Coenonympha tullia*) have declined at the southern edges of their ranges more rapidly than further north. The garden tiger moth (*Arctia caja*), a spectacular insect common everywhere 50 years ago, has declined by as much as 90 per cent since then and has survived best at higher, cooler elevations. In Somerset, that includes the Mendip Hills, but at lower levels, including in our garden, it is now rarely seen. Four of Britain's dragonflies/damselflies are

Figure 6.2. Snowdon lily, an Arctic–alpine plant declining due to climate warming. © Paul Sterry/Nature Photographers Ltd (A black and white version of this figure will appear in some formats. For the colour version, please refer to the plate section.)

predominantly northern species. Azure hawkers (*Aeshna caerulea*), northern emeralds (*Somatochlora arctica*), northern damselflies (*Coenagrion hastulatum*) and white-faced darters (*Leucorrhinia dubia*) have all retreated northwards. And bilberry bumble bees (*Bombus monticola*), denizens of upland habitats, have withdrawn along their southern distribution limits in northern England. Lepidoptera and Odonata are the most visible and well-studied invertebrate groups, and very likely there have been others impacted adversely by climate warming that have remained largely out of human sight.

Arguably, the most severe downside of climate change has been felt in our coastal marine ecosystems, especially the North Sea. Coastal waters around the UK have mirrored terrestrial habitats by warming up continuously since the 1980s. This change has enticed fish, mammals and invertebrates previously of more southerly distributions to turn up in Britain, especially along southern and western approaches. Sardines (*Sardina pilchardus*) and red mullet (*Mullus surmuletus*) have advanced into the North Sea. Flat topshells (*Gibbula umbilicalis*) and toothed topshells (*Osilinus lineatus*) have progressed eastwards along the English Channel coast, and several cetaceans including Blainville's beaked whale

(*Mesoplodon densirostris*) are increasingly common visitors. These are perhaps welcome additions to the UK's marine fauna, but they hardly compensate for what has happened to the North Sea ecosystem. The average temperature of this waterway has risen by 1°C since the 1970s with dramatic consequences for the food web it supports. In past times, a spring bloom of phytoplankton precipitated an efflorescence of zooplankton predators, dominated by the copepod (*Calanus finmarchicus*). This invertebrate in turn provided prey for sand eels, huge shoals of which supported seabird chicks in nurseries on cliffs around the British seaboard. Some of these seabird colonies are enormous, running into many thousands, providing wonderful spectacles set in dramatically attractive coastal scenery. Unfortunately, *Calamus finmarchius* is intolerant of the warmer temperature regime, and has been largely replaced by the southerly copepod (*Calamus helgolandicus*). This matters, because the newcomer does not bloom until later in the year, which is not in time for the sand eels to benefit in spring. The upshot has been fewer eels at a critical time of year and therefore less food for seabird chicks. Some coastal colonies of our hugely impressive cliff-nesting fauna have declined as a result. Kittiwakes have been particularly hard hit, relying heavily as they do on the sand eel crop. Kittiwakes are not alone in this crisis, and downturns of some charismatic puffin, guillemot (*Uria aalge*) and razorbill (*Alca torda*) populations are likely due to shortages of this key prey item. The knock-on effects of climate change in the North Sea have been among the most serious deleterious impacts of a warming world on British wildlife.

Climate change is a global issue and, unlike problems described in earlier chapters, it cannot be ascribed solely, let alone mostly, to what goes on in the UK. However, British efforts to limit global warming are not unrelated to human numbers on the islands, as will be discussed later.

6.3 Disease and Wildlife Declines

Pathogens and Hosts

Pathogens are the constant companions of plants and animals in all ecosystems. The evolutionary arms race between hosts and infectious agents, usually bacteria, viruses, fungi or protozoans, is similar in principle to that between predator and prey. In a 'closed' system, with little or no immigration or emigration, a cyclical pattern of host population change is commonplace. Declines follow a round of infection, during which resistant individuals are selected for and host immunity becomes widespread. Population recovery persists until new mutations able to evade the immune response arise in the pathogen, after which the cycle goes round again. In such a system, host extinction is expected to be rare, because modification of the immune response to cope with an organism not vastly different from what came before is usually rather fast and effective.

The big change in this situation within historical times has been a shift from 'closed' to 'open' systems, facilitated by movement of pathogens well beyond their regions of origin where hosts were routinely rescued by immune response cycles. Most of this increased disease mobility has been caused inadvertently by *Homo sapiens*, a nomadic species ever on the move since our diaspora from Africa tens of thousands of years ago. From very early on, there is evidence from cave art and artefacts that even when movement was solely on foot, traded items and customs travelled many thousands of kilometres across Eurasia. But human movements increased dramatically with the sequential inventions of horse-drawn carts, ocean-going ships, motorised vehicles and aircraft, with ominous consequences for disease transmission.

Humans were among the first species to suffer the consequences of pathogens travelling along with them. Plagues, often poorly documented, have long been present and were dreaded in ancient times, as reported in the Bible among other sources. But it was in the mediaeval period that a truly dreadful pathogen was unleashed and recorded clearly enough that its cause and origins would eventually be unveiled. The Black Death, mediated by the bacterium *Yersinia pestis*, spread like wildfire across Europe between 1348 and 1351. Estimates of the death toll are inevitably uncertain, but probably at least 25 million people – perhaps a third of the population – died in Europe at that time. That was not the end of the matter, because although resistance developed among the survivors, there were future severe outbreaks, such as one in London that killed 70,000 people in 1665–1966. This bug is still around and has infected people in the USA and Asia in recent times, but fortunately is easily treated with antibiotics. The key to the original devastation was its movement from an ancestral home in central Asia. People travelling westwards to the Mediterranean arrived with infected black rats together with their fleas, all of which thrived in a novel environment and donated *Y. pestris* to humans previously naïve to these newcomers. This kind of history has repeated itself many times over. An especially awful example followed the discovery and exploration of the Americas by European sailors, starting in the sixteenth century. Only recently has it come to light that a great civilisation of many millions of people lived in the Amazon basin, which was not entirely an ancient pristine rainforest as generally believed. Perhaps 90 per cent of the native inhabitants were wiped out by diseases such as measles and influenza, carried in by European visitors largely immune to pathogens that had coevolved in their home continent for thousands of years. There are all too many more examples of translocated pathogens wreaking havoc on naïve human populations, probably assisting the colonisation by Europeans of places as far apart as North America and Tasmania. Empire building relied at least as much on microbes and viruses as on firepower. And it goes on, with intermittent influenza outbreaks, and with the coronavirus epidemic of 2020 showcasing the rate at which newly emerged pathogens can benefit from air travel and spread almost overnight across the entire globe.

Another source of novel human pathogens is wildlife itself. Zoonosis, defined as a disease transferred between different species, has become another serious problem for humans. Some types of influenza evolved in birds, especially domesticated wildfowl, and successfully infected humans often after genetic recombination with human viruses. This has led to serious disease outbreaks, but two zoonoses originating in Africa have been especially ominous products of twentieth-century interactions with wildlife. Human immunodeficiency virus (HIV) transferred from either chimpanzees or gorillas (*Gorilla* spp.) into humans where it causes acquired immune deficiency syndrome (AIDS), which, until treatments became effective, was usually fatal. Similarly, the dreadfully devastating Ebola virus almost certainly originated in bats somewhere in central Africa. Probably the bush meat trade contributed to these cross-infections, which require contact with body fluids. And it seems that coronavirus pathogens also originated in wildlife, probably again in bats but via an intermediate host, possibly pangolins (species uncertain), in Chinese wet markets. Pinpointing the exact lines of cross-species disease transfers is difficult and relies heavily on inferences from DNA sequences, but there is no doubt that it has happened in at least a few well-documented cases.

The severity of infectious disease outbreaks is, in almost all cases, related to the population density of the host organism. Transmission is most efficient in a crowd, and it is no coincidence that pandemics among humans strike hardest in towns and cities where many people live or work in close proximity.

Plant Diseases in the UK

There is no shortage of plant pathogens in nature. As an undergraduate in the 1960s, I recall university courses devoted to these organisms, many of which cause serious damage to crops and garden plants. According to the Royal Horticultural Society (RHS; www.rhs.org.uk) the most problematic diseases in 2019 were:

1. Honey fungus (*Armillaria* spp.), a native organism that causes root rot in many perennial plants.
2. *Phytophthora* spp., especially *Phytophthora ramorum,* a water mould that thrives in damp conditions. First seen in Britain in 2002 and now widespread around the world thanks to the plant trade, it has devastated larch (*Larix decidua*) plantations to the extent that this tree can no longer be used for timber in Britain. The related *Phytophthora infestans* arrived much earlier, probably from Mexico, and was the infective agent of the Irish potato blight. This disease outbreak was responsible for the dreadful famine and countless thousands of deaths in that country in the mid-nineteenth century.
3. Box blight (*Cylindrocladium buxicola*), a fungal pathogen primarily of box (*Buxus* spp.), browning leaves and causing dieback of young stems. Box trees and shrubs are locally common, and Box Hill in Surrey is named on account of the box woodland there. The disease was first noticed in the 1990s in Britain, but its origin is unknown.
4. Pear rust (*Gymnosporangium sabinae*), yet another fungus, requiring both pear and juniper trees to complete its life cycle. Bright orange spots on the upper surfaces of leaves are diagnostic of pear rust, which weakens rather than kills its host. Until recently, this pathogen was rare in Britain, and was mostly confined to mainland Europe.
5. Brown rot fungus (*Monilinia laxa* and *M. fructigena*), causing widespread damage to fruit trees, including apples, pears, plums and cherries. The fruit rots away, and in spring, the same fungus kills blossom and small branches.
6. Leaf spot and canker of *Prunus* trees. Bacterial canker caused by *Pseudomonas syringae* can damage or kill *Prunus* spp., especially plum and cherry trees. The microbe creates dead patches of bark (cankers) and small holes in leaves.
7. Apple and pear scab, two closely related fungal diseases (*Venturia* spp.) that cause dark, scabby marks on the fruit and leaves of apples and pears.

8. Rose black spot, caused by the fungus *Diplocarpon rosae*. This is all too familiar to gardeners and generates dark spots on rose leaves, which drop off early, substantially weakening the host plant.
9. Powdery mildew, infecting a wide range of plants, covering the leaves with whitish spots. Fungi in the genus *Erysiphales* are the agents responsible, eventually reducing the vigour of its hosts.

The RHS focuses on species of horticultural interest, but of course there are very many more pathogens that attack wild plants. Notable aspects of the RHS list are the predominance of fungi as infective agents, and recognition that several of the causative microbes are relatively new to Britain. Not mentioned are two disease outbreaks that have had very substantial impacts on the British countryside, notably Dutch elm disease (DED) and ash dieback (AD). Until the mid-twentieth century, elm trees (English elm, *Ulmus procera*, and wych elm, *U. glabra*) dominated countless English woodlands. Many birds eat elm seeds and the leaves provide food for caterpillars, including, in the case of wych elm, those of the white-letter hairstreak butterfly (*Satyrium w-album*), an insect that has declined dramatically since DED arrived in the UK. This disease, caused by the fungus *Ophiostoma novo-ulmi* and spread by the otherwise innocuous bark beetle (*Scolytus scolytus*), has killed millions of English and wych elms since the 1960s. The apparently pejorative name of DED stems not from its country of origin but from a group of Dutch scientists who identified it in the 1920s. Before this pandemic started, there were an estimated 30 million elm trees in Britain, with about half of these in southern England. The fungus is not native to the UK, and its most virulent form was imported accidentally from North America in the early 1960s (Potter et al., 2011). Attempts to control the spread of DED by removal of infected trees have mostly failed. In Sussex during the 1970s, there were road signs announcing entry to a DED area, although it was never clear as to how passers-by were expected to respond to them. However, the University of Sussex, founded more or less contemporaneously with the arrival of DED, has spent a lot of time and money attempting to preserve an impressive avenue of elms running along the campus's central valley. Although occasional infected trees are still removed, the conservation efforts so far have been pleasingly successful. Elsewhere, elms still survive in hedgerows because roots are not killed by DED and regenerate root sprouts ('suckers') after the main tree dies, although these rarely grow more than 5 m tall before succumbing to a renewed attack of the fungus.

A more recent assault on Britain's woodlands has followed the arrival of another fungus, *Hymenoscyphus fraxineus*, originally native to eastern Asia but which has spread westwards, arriving in Europe in the 1990s and in England in 2012. It has subsequently decimated ash trees (*Fraxinus excelsior*) across the continent and AD is changing the landscape over much of the UK (Figure 6.3). It is expected that around 80 per cent of Britain's ash trees will eventually be killed, although fortunately some individuals are resistant to fungal attack and these can be propagated to (hopefully) repopulate the country's extensive ash woodlands in the long term (Evans, 2019). On the Mendip Hills, where AD is rampant at the time of writing, die-offs are particularly dramatic and removal of dead wood is transforming the landscape in well-known beauty spots such as the Cheddar Gorge. AD is spread as wind-blown spores, which might have arrived in the UK without human help, but the disease was also brought in accidentally from abroad with ash saplings.

Many of the most devastating plant diseases in Britain are relatively new arrivals from foreign parts. This situation, largely a consequence of increasing movements of people around the globe, has thus far proved impossible to regulate. Lack of effective controls on plant imports, and therefore of the pathogens they carry with them, seems likely to sustain this problem well into the future.

Figure 6.3. Ash dieback. *Source*: Bethany Clarke/Stringer/Getty Images

Animal Diseases in the UK

Newly arrived alien pathogens have caused plenty of problems for British animals in recent decades. This is despite that fact that legislation supposedly restricting imports of non-native species lacks a loophole comparable with that enjoyed by horticulturalists, which allows largely unfettered entry of foreign plants into the UK.

Probably the best-known disease of wild animals in Britain is myxomatosis, caused by myxoma virus, originating in the Americas where it causes a mild disease in hosts such brush rabbits (*Sylvilagus bachmani*) with which it coevolved. In European rabbits, the outcome of infection is very different. Before its arrival in the UK in 1953, almost certainly by deliberate introduction, the rabbit population was estimated to be around 60–100 million (Bartrip, 2008). The European animal turned out to be highly susceptible to myxomatosis, with initial mortality rates of at least 99.5 per cent. Within a few years, the disease spread across the whole of the UK, and by 1957, only between 1 and 6 million rabbits remained. Farmers' delight at the overnight annihilation of a major agricultural pest was not matched in the wider community. Many people were dismayed by the disappearance of 'Peter Rabbit', and Prime Minister Winston Churchill fumed against myxomatosis, attempting (unsuccessfully) to make its deliberate spread illegal. Over the following 50 years, disease resistance of the survivors spread and the rabbit population rebounded to perhaps 40 million by 2004. But for this unlucky species, more bad news was to come. Rabbit haemorrhagic disease virus (RHDV), a calicivirus of the genus *Lagovirus,* which probably originated in China, arrived in Britain in the 1990s. This virus is genetically variable, and there have been at least two major outbreaks in Britain with 'type 2' first detected in the UK in 2013. Mortality rates from RHDV are high, and rabbits are once again disappearing from many parts of the British countryside. They no longer visit our garden, and in our part of Somerset, just sighting one has become a notable event. Alarmingly, RHDV and possibly myxomatosis have spread into brown hare populations, causing extra mortality in a species that has already declined substantially due to agricultural intensification. The consequences of rabbit declines extend far beyond the species itself, and include alterations in floral and invertebrate diversity as well as scrub encroachment in rare habitats. Rampant invasion of sea buckthorn (*Hippophae rhamnoides*) on the Sefton coast dunes in the late twentieth century began following the disappearance of rabbits in the 1960s. Vast areas of previously open dune habitat with its specialised flora and fauna have been overwhelmed by impenetrable thickets of the invading shrub, leading to widespread declines of local plant and animal rarities.

Neither rabbits nor hares are native to the UK, although both species have been part of the country's ecology for many centuries. Other new diseases, however, have decimated some long-standing natives. Red squirrel declines followed the introduction of North American greys in the late nineteenth century. Partly, this may be down to the larger size and thus physical dominance of the newcomer, but, analogous to the situation with myxomatosis, grey squirrels harbour a squirrelpox virus, which to them is innocuous but which to red squirrels is usually fatal. This infection has exacerbated, possibly even driven, the impact of the invader on the native species (Chantrey et al., 2014). And disease, in this case the paramyxovirus phocine morbillivirus (formerly phocine distemper virus), has caused mass mortalities of harbour seals (*Phoca vitulina*) in European waters in outbreaks during 1988 and 2002. In this instance, the disease origin is unknown but is suspected to have originated in Arctic seal populations.

Ornithologists are increasingly aware of sicknesses in wild birds, including visitors to garden feeders. Greenfinches (*Chloris chloris*) were among the first, and remain the main, victims of trichomonosis disease. The protozoan *Trichomonas gallinae* causes breathing difficulties, and infected birds become lethargic with puffed-up feathers, eventually dying. The results of an epidemic of trichomonosis have been substantial declines of vulnerable species, mostly finches, first noted in 2006 and ongoing in much of the country. It probably originated in wood pigeons, although this is not known for certain. Another bird much in the news, following its virtual disappearance in places where this well-loved cohabiter with humans previously prospered, is the house sparrow. Declines of this bird have been most marked in cities, and a study in London revealed damaging effects of a bird malaria parasite, *Plasmodium relictum* (Dadam et al., 2019). Survival of young sparrows correlated inversely with *Plasmodium* levels in their blood, so this protozoan may have contributed significantly to population declines. Just as with human malaria, the avian disease is transmitted by mosquitoes, and global warming, which correlates with increased bird infection rates in temperate countries, may have increased the abundance of these vectors.

Among vertebrates, amphibians hold the unenviable record of experiencing the highest rates of declines and extinctions in the world. By the early 2000s, it was realised that perhaps a third of all amphibians were undergoing serious declines or were already extinct (Beebee & Griffiths,

2005). Permanent losses included some high-profile species, including the golden toad (*Incilius periglenes*) of Costa Rica and the stomach-brooding frogs (*Rheobatrachus* spp.) of Australia. As with other wildlife, habitat destruction has been responsible for most of the damage to amphibians, but a new and especially virulent disease has also made its mark. The fungal pathogen *Batrachochytrium dendrobatidis*, discovered in the late 1990s, causes chytridiomycosis. This disease affects the function ing of amphibian skin, causing rapid mortality, and has been responsible for widespread population declines, especially in tropical countries. Genetic studies show it originated in eastern Asia (Fisher et al., 2009) and was presumably spread elsewhere by human agency, especially via a minimally regulated pet trade. *B. dendrobatidis* arrived in Britain in the early 2000s and quickly became widespread among the native amphibians. Fortunately, it has proved benign in the UK, although it is carried passively at substantial levels in some species. Of greater significance to British amphibians have been recurring outbreaks of ranavirus infections, especially in common frogs. This infection causes widespread mortality events, wiping out as much as 80 per cent of a local population and leaving a collection of gruesome carcasses (Figure 6.4). The virus is thought to have originated in North America and may have been transported with goldfish or other amphibians for the pet trade. Outbreaks are most frequent in southern and eastern England, providing unpleasant spectacles for garden pond owners. Since the 1980s, ranavirus outbreaks have spread northwards and eastwards, and are strongly correlated with human population density (Price et al., 2016). This relationship is no doubt due to the high numbers of garden ponds in urban areas, which are very popular breeding sites for common frogs, in contrast to the relative scarcity of suitable ponds in the farmed countryside.

Invertebrates have also suffered from disease-mediated die-offs. However, although this class of animals accounts for the overwhelming majority of biodiversity in the UK, precious little is known about the ecology, let alone disease issues, of most of the thousands of species inhabiting these islands. Who would notice an epidemic among millipedes? There are, however, some invertebrates in which the effects of pathogens are well documented. The disappearance of the native white-clawed crayfish in so many of Britain's streams and rivers has unhappy parallels with the fate of red squirrels. The introduced American signal crayfish is larger than the native species, but its impact has been mediated largely by the pathogenic water mould *Aphanomyces astaci* for which signal crayfish are passive carriers. Unfortunately, white-clawed crayfish

Figure 6.4. Common frog killed by ranavirus. Photograph by Jim Foster

are highly susceptible to 'crayfish plague', a disease caused by the mould. Entire populations rapidly succumb when it invades their freshwater habitats. As with red squirrels, conservationists are engaged in an epic struggle to prevent the native species from becoming extinct. At current rates of decline, the UK may lose this noble crustacean within a couple of decades.

One particular group of well-studied invertebrates also has a disease story to tell. Honey bees, the semi-domesticated descendant of a European wild bee, are reared in enormous numbers for commercial pollination as well as on smaller scales in garden hives. Given the high densities of these insects kept in industrial-scale apiaries, it is perhaps unsurprising that mortality from diseases can be high. Parasitic *Varroa* spp. mites spread westwards from Asia and appeared in England in 1992. They have caused the demise of many hives and have proved impossible to eradicate, although some control methods have been deployed successfully. However, pollination of flowering plants, including crops such as oil seed rape, depends much more on wild bee species and other insects than it does on honey bees. An ongoing crisis of declines in many of these bees, precipitated largely by pesticide applications (see Chapter 5), has attracted increasing attention in recent years. Bumble bees, in particular, are among these vital pollinators, and are a summer delight as they buzz busily in gardens across the country. Some species have, like honey bees, been exploited commercially, especially for pollination of crops such as tomatoes, which they service very efficiently. Buff-tailed bumble bees (*Bombus terrestris*), in particular, are reared in great quantity (millions of nests) and have been exported to many countries around the world. As outlined by Goulson (2013), this is where disease risks arise. Escaped buff-tails have established colonies in foreign climes, taking with them parasites such as the trypanosome protozoan *Crithidia bombi*. Although this organism can cause disease in European bees, it is not usually devastating. However, in South America, the spread of buff-tails has been linked to the decline of one of the world's largest bumblebees, the 'flying mouse' (*Bombus dahlbomii*). Although the evidence is circumstantial, declines of the flying mouse are thought to be down to the co-invading trypanosome. There are other suspicious declines of native bee species where buff-tails have been introduced, so although disease has not had a major impact on British bumble bees, it seems that their export has been seriously damaging elsewhere in the world.

It is clear that disease has played an increasingly prominent role in wildlife declines over recent decades, although not at a level comparable with habitat destruction. It is also evident that a substantial number of the pathogens responsible are newly arrived in the UK, mediated accidentally in most but not all cases by ever more widely travelled humans.

6.4 Damage by Disturbance

Disturbance as an Issue

One activity that has only been considered rather recently is regular disturbance and its possible consequences for wildlife. Many facets of behaviour are critical to the survival and successful reproduction of wild animals. These include feeding, drinking, access to refugia, courting, mating, production and sometimes the care of progeny, and maintaining an appropriate body temperature. All of these aspects of life must be accomplished in an environment that is as safe as possible from the attention of predators. Most species live in constant danger of attack, and react rapidly to any perceived threat. This is the context in which unintentional disturbance by humans could, if severe or frequent, cause sufficient stress to affect population viability. Just watching garden bird feeders gives an indication of how disturbance works. The flocks of goldfinches we see on ours every day have never become conditioned against taking flight when people emerge from a nearby door, despite the fact that they have witnessed this event for months on end. Much energy is expended wheeling around and re-alighting on the feeders just minutes later. How much more alarm must human intrusion cause in wild places where people have not provided incentives for animals to become peacefully acquainted with them?

There seems little doubt that in some situations disturbance of wildlife by humans is severe and has been increasing. Visitor numbers to attractive venues such as the English Lake District National Park have risen steadily for decades. In 2018, over 18 million tourists enjoyed the glorious mountain and lake scenery in this part of Cumbria, up from around 15 million in 2012. The park authority in 2018 anticipated a 5 per cent per year increase in visitor numbers, up to maybe 46 million by 2035. Other popular places are faring similarly, and many, like the Lake District, are home to some of the UK's most precious wildlife species. The most obvious damage from so many visitors is often the degradation of footpaths, including the destruction of surrounding vegetation. In the Dartmoor National Park and on the South West Coast Path in south-west England, the average soil loss for paths has been estimated at 1.41 kg per metre per year, whereas 'off-path' soil loss, where natural soil redistribution processes occur, was just 0.79 kg per metre per year (Rodway-Dyer & Walling, 2010). This level of damage necessitates continuous maintenance, often carried out by volunteers, and looks unsustainable if pressures continue to increase.

Increasing numbers of dogs, especially those off the lead, are also of escalating concern. Phil Smith (2017), an experienced naturalist and long-standing documenter of wildlife on the Sefton dunes, documented the impacts of dogs on this important wildlife area and elsewhere. Damage has included disturbance and predation of ground-nesting birds, attacks on livestock in conservation grazing areas, and rampaging uncontrolled through shallow dune slacks. Smith also points out more general problems with a large population of dogs, noting that because of what dogs eat, the carbon 'footprint' of a medium-sized dog is about twice that of a large four-wheel-drive vehicle. It takes an estimated 0.84 ha of farmland to feed a dog, while the vehicle uses the energy equivalent of 0.41 ha of land. It is doubtful whether this is sustainable in the long term.

Disturbance of Wild Birds

Probably because they are conspicuous and widely watched members of British wildlife, there are more observations concerning the effects of disturbance on birds than on any other animal group. An early study in Finland showed that velvet scoter (*Melanitta fusca*) ducklings alarmed by

Figure 1.3. Wildfires raging in California. *Source*: Kent Nishimura/Los Angeles Times/Getty Images (A black and white version of this figure will appear in some formats.)

Figure 2.3. House martins, one of many species declining in England. *Source*: Arterra/Universal Images Group/Getty Images (A black and white version of this figure will appear in some formats.)

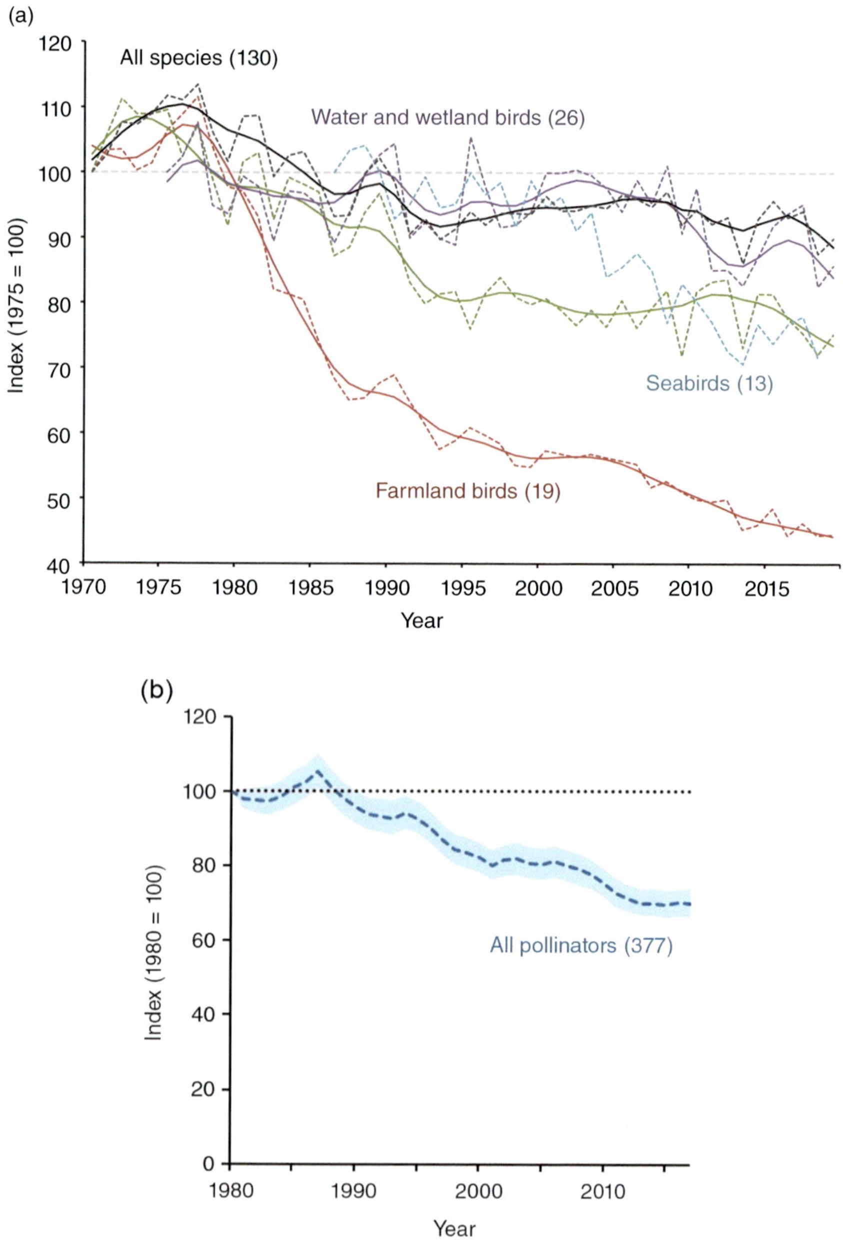

Figure 2.4. (a) Changes in the abundance of breeding birds of woodland, farmland, water and wetlands, and all species in the UK. *Source*: BTO, DEFRA, JNCC, RSPB. (b) Line graph showing indicator trend of 377 pollinator species (wild bees and hoverflies) declines in the UK, with shaded 90 per cent confidence limits. *Source*: Defra 2020. UK Biodiversity Indicators. Indicator D1c. Status of pollinating insects. Scheme; Source: Bees, Wasps & Ants Recording Society; Biological Records Centre (supported by UK Centre for Ecology & Hydrology and Joint Nature Conservation Committee); hoverfly recording (A black and white version of this figure will appear in some formats.)

Figure 3.1. Eurasian wolf, the last native mammal exterminated in the UK by human persecution. *Source*: Ginger/Moment/Getty Images (A black and white version of this figure will appear in some formats.)

Figure 3.2. Summer lady's tresses, exterminated in Britain by specimen collection. Photograph by Hans Dekker, reproduced with permission from Saxifraga Foundation (A black and white version of this figure will appear in some formats.)

Figure 4.2. Natterjack toad. (A black and white version of this figure will appear in some formats.)

Figure 5.3. Wildflower meadow, once but no longer a common sight on UK farms. *Source*: Ashley Cooper/The Image Bank/Getty Images (A black and white version of this figure will appear in some formats.)

Figure 5.4. Terminal decline of a farm pond. Photograph by Joe Costley/Plantlife, UK (A black and white version of this figure will appear in some formats.)

(a)

(b)

Figure 5.5. Damage done by bottom trawling. (A) Before trawling. (B) After trawling. *Source*: Colin Munro, Marine Bio-images/Colin Munro Photography (A black and white version of this figure will appear in some formats.)

Figure 5.6. Wareham Forest, once an extensive open heath. *Source*: Simon Byrne/ RooM/Getty Images (A black and white version of this figure will appear in some formats.)

Figure 6.2. Snowdon lily, an Arctic–alpine plant declining due to climate warming. © Paul Sterry/Nature Photographers Ltd (A black and white version of this figure will appear in some formats.)

Figure 6.5. Male adders wrestling at breeding time. *Source*: Education Images/ Universal Images Group/Getty Images (A black and white version of this figure will appear in some formats.)

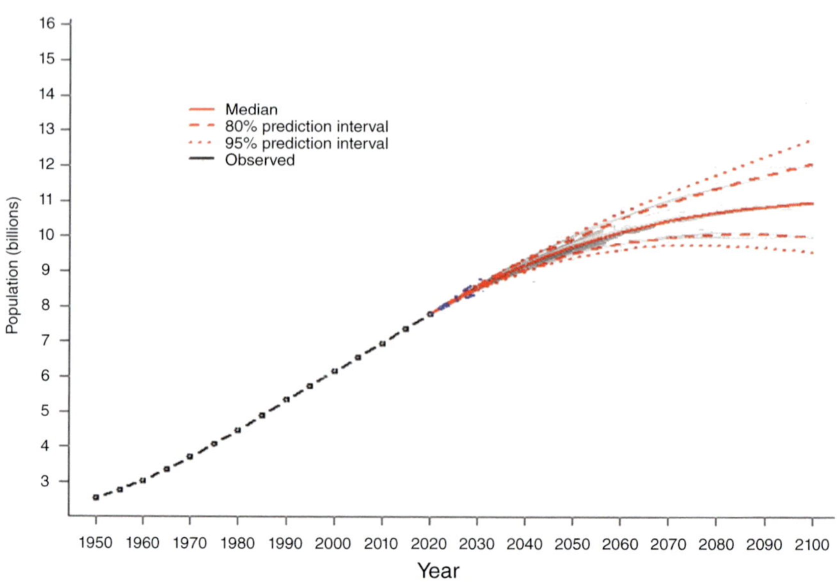

Figure 7.2. Projections of future total world human population trends. From World Population Prospects 2019, http://population.un.org/wpp. © 2022 United Nations. Reprinted with the permission of the United Nations (A black and white version of this figure will appear in some formats.)

Figure 8.1. Beavers are back, to the delight of many people in the UK. *Source*: Arterra/Universal Images Group/ Getty Images (A black and white version of this figure will appear in some formats.)

Figure 8.4. The case for halting eternal economic growth. *Source*: Meteor Independent Media, https://themeteor.org/category/editorial/ (A black and white version of this figure will appear in some formats.)

Figure 9.1. A distinct border between the Serengeti reserve (left) and agricultural land (right). © Daniel Rosengren (A black and white version of this figure will appear in some formats.)

Figure 9.3. The UN Sustainable Development Goals. *Source*: UN Sustainable Development Goals, www.un.org/sustainabledevelopment/. The content of this publication has not been approved by the United Nations and does not reflect the views of the United Nations or its officials or Member States (A black and white version of this figure will appear in some formats.)

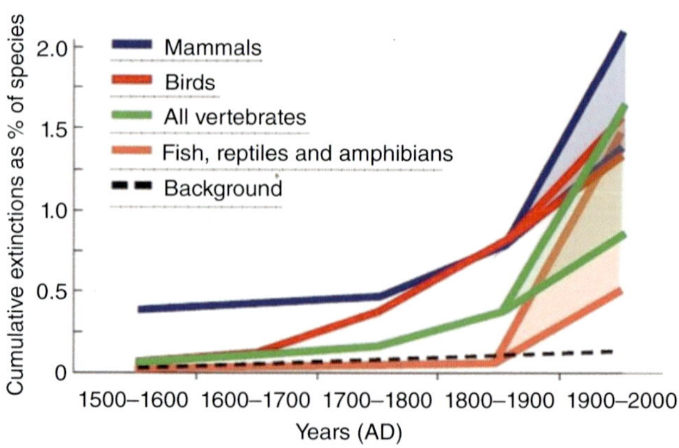

Figure 9.4. Patterns of animal extinctions over time. Permission courtesy of British Geological Survey © UKRI. *Source*: https://science.sciencemag.org/content/351/6269/AAD2622.abstract (A black and white version of this figure will appear in some formats.)

Figure 10.3. Amphibian and Reptile Conservation (ARC) volunteers carrying out heathland management work in Surrey. © Bryony Davison, ARC (A black and white version of this figure will appear in some formats.)

passing leisure boats swam further, with less feeding time, than undisturbed chicks. The frequency of attacks by gulls (*Larus argentatus* and *L. marinus*), the main predators of velvet scoter ducklings, was more than three times higher for disturbed compared with undisturbed birds (Mikola et al., 1994). This kind of disruption is common. Boating on inland waterways in the UK has enjoyed a boom time in recent years, and environmental damage has been a significant consequence. In the Norfolk Broads, many of the connecting channels between the lakes are continuously churned up, limiting the establishment of submerged plants and rendering the waterways visually unattractive. Restoration of the Basingstoke Canal in south-east England opened up for boating a previously biodiverse habitat that, in the 1970s, boasted dense growths of aquatic plants as well as large colonies of common frogs and toads. Much of this wildlife has disappeared, victims, as in the Broads, of conversion of the canal into a muddy ditch.

Some coastal birds, especially those inhabiting mudflats or nesting on beaches, are also experiencing increasing visitor pressure. Sometimes this does not seem to matter. Detailed observations of 10 species frequenting the Wash mudflats concluded that extra flights induced by humans were unlikely to impact significantly on their foraging times. Further east along the Norfolk coast, the situation was rather different. Visitor pressure at Holkham has strongly restricted the distribution of ringed plovers on the beaches, and the few remaining areas of suitable habitat may disappear if tourism levels increase (Coombes et al., 2008). Away from the coast, some species of farmland birds in the Netherlands, including lapwings (*Vanellus vanellus*) and black-tailed godwits (*Limosa limosa*), have responded to disturbance by nesting at least 1 km away from villages and major roads. Perhaps surprisingly, blue tits can also respond negatively to stresses from disturbance caused by human activity. In the UK, these birds are among the commonest in garden habitats, where they forage and nest in considerable numbers. They appear very tolerant of people around them, including researchers who investigate their nests in artificial boxes. In Spain, however, nestlings in recreational areas with regular public pressure grew more slowly if they hatched in holiday periods, when many people were visiting, compared with those that hatched in quieter times. Furthermore, the disturbed offspring fledged with lower body weight and in poorer condition than those emerging outside holiday times, or in local woodlands. These differences may well translate into variations in subsequent juvenile survival (Remacha et al., 2016). It is a particular worry that some birds that react adversely to disturbance in Britain are relatively rare. Nightjars

(*Caprimulgus europaeus*), a summer visitor in northern Europe, are a priority species within the UK Biodiversity Action Plan and are on the Red List of species of conservation concern. Between 1972 and 1992, the nightjar's breeding range decreased by more than 50 per cent due to loss or degradation of its heathland breeding habitat. At a Nottinghamshire site of mixed heath and woodland, nightjars declined by about 10 per cent over a decade, and their density was lower, with fewer breeding pairs, in heavily disturbed compared with less disturbed habitat areas (Lowe et al., 2014). Although average breeding success per pair was not significantly different between the two habitats during the study years, human recreational disturbance drastically altered the settlement patterns and nest site selection of arriving females. This is a ground-nesting species that is extraordinarily well camouflaged when incubating eggs, making it highly vulnerable to people approaching without seeing the bird until almost on top of it. Management clearly needs, wherever possible, to protect nesting areas for this species from human footfall.

Disturbance of Other Wildlife

Among mammals, cetaceans and seals have attracted attention as possible casualties of excessive disturbance by human activities out at sea. Two main issues have come under scrutiny. Whale and seal watching have become increasingly popular, while interactions of cetaceans with large vessels include physical damage to the animals and interference with their underwater communications by low-frequency sound emissions. Whale watchers approach whales and dolphins, and also basking sharks (*Cetorhinus maximus*), in small boats that have the potential to stress the animals unless care is taken to avoid close proximity and lengthy encounters. Boats can disrupt the feeding behaviour of these animals, potentially reducing fitness and posing a threat to small populations. There are guidelines for best practice in this kind of tourism that are followed by most operators, but in Scotland these are based on several voluntary codes of conduct that are not always mutually consistent (Inman et al, 2016). Greater uniformity of advice is clearly needed. Observations of harbour porpoise (*Phocoena phocoena*) responses to passing ships and boats have revealed negative reactions (moving away or diving) to some, mostly high speed, vessels. Whether any of these interactions have significant consequences for the long-term viability of cetacean populations is unknown, but application of precautionary principles attempting to minimise interference with these impressive creatures is surely the safest way forward.

In the case of terrestrial mammals, most are too secretive for serious disturbance studies. Comparisons of mammal assemblages including 14 species, ranging in size from voles to deer, in quiet and disturbed areas of Epping Forest found no significant differences between the two regions. However, the culling operations inflicted on badgers in attempts to control the spread of bTB have generated incidental experiments on the effects of such disturbance on these mustelids. Common responses have been emigration of some survivors out of the cull areas, and increasing ranges of indigenous and surrounding social groups (Tuyttens et al., 2000). This behavioural response of badgers to predation pressure has complicated attempts to evaluate the effectiveness of the cull strategy on bTB control. Other mammals at risk from disturbance, especially during winter hibernation, are bats in house roof spaces and in caves popular with tourists. Legislation, in theory at least, protects bats from all damaging activities, and because several species seem to be improving their lot in Britain, the law is perhaps working reasonably well.

Reptiles have a particular susceptibility to disturbance on account of their need to thermoregulate, a behaviour that typically involves basking in sunshine. This activity is commonplace in spring after emergence from hibernation, and the animals can be very conspicuous at this time, with mating behaviour making them very exposed (Figure 6.5). In the UK, adders have declined a lot in recent years due to multiple factors including habitat destruction. However, monitoring results from 260 adder sites across the country since 2005 has highlighted disturbance as the main perceived problem for the snakes (Gardner et al., 2019). Public pressure was manifest as dog walking, mountain biking, photography, passing vehicles, vegetation trampling and even collection of the reptiles. In Somerset, there are habitats widely known to hold adder populations, and these attract many photographers every spring. Other reptiles may suffer similarly, but most species are rarer or less conspicuous than Britain's only venomous reptile.

Finally, some insect populations can also become casualties of human interference. Bumble bee nests are destroyed by several agents, among which badgers probably account for most losses by predation, but humans, by gardening or via construction work, seem to come second. The fate of pollinators, like so many insects, is nevertheless predominantly dictated by habitat degradation in the wider countryside.

Figure 6.5. Male adders wrestling at breeding time. *Source*: Education Images/ Universal Images Group/Getty Images (A black and white version of this figure will appear in some formats. For the colour version, please refer to the plate section.)

6.5 Overview

Although climate change, disease and disturbance have clearly affected wildlife in Britain, their combined impacts have been substantially less than factors described in earlier chapters, especially agricultural intensification. Even so, the question arises as to how much these subsidiary influences relate to the human population size.

Climate change is driven by the emissions of greenhouse gases, which are produced all over the world. Pollution is not evenly spread, with 20 countries accounting for about 80 per cent of the total. In 2020, the UK came 17th in a league table of gas emitters but 14th on a per-capita basis. Unsurprisingly, the top 20 polluters excluded most developing countries, including almost all of Africa and of the Americas south of

Mexico. Affluent states such as the UK are inflicting climate change on poorer countries that, in many instances, are suffering the most serious effects of it. The British per-capita performance is relatively good compared with some others in the top 20, and heading in the right (downwards) direction. As individuals, we are entreated to take a range of measures to reduce our 'carbon footprint', including eating less meat, flying less often and improving our house insulation. Thinking twice about dog ownership could be a useful addition to this list. These are important actions, but a comparative study showed that, in the long term, a decision to have one fewer child by families in industrialised countries would be 20 times more effective in reducing CO_2 emissions than any of the measures listed above (Wynes & Nicholas, 2017). Population size is evidently an important contributor to climate change, albeit one that can only be controlled in the future rather than immediately.

Outbreaks of new wildlife diseases have increased since the Second World War, almost certainly due to human-mediated transfers. In this case, there is little reason to believe that the concurrently rising human population has been responsible. A much more likely cause is the explosion in cheap air travel and the greater movement of people and animals around the world, some as unsuspecting vectors of damaging pathogens. The incidence of ranavirus infections in English frogs correlates with human population density, but this most likely reflects a corresponding correlation with the numbers of garden ponds in semi-urban areas. More people means more ponds, which translates into more frog populations open to virus infection. Only improved biosecurity measures hold out a hope of reducing the invasion of ever more pathogens into Britain, but the coronavirus pandemic of 2020 is a stark example of how difficult that can be, even with diseases that attack humans.

Disturbance is a problem that seems inherently likely to depend on human numbers, but even in this case, any relationship with population size is problematic. As restrictions relating to coronavirus were temporarily relaxed during the summer of 2020, people took the opportunity to invade the UK countryside in unprecedented numbers. In popular sites, visitor numbers rose by at least 40 per cent compared with the previous year, leading to widespread damage in some of the country's most beautiful and ecologically important habitats. Mountains of litter, and heath fires following discarded barbeques, featured strongly in that summer's legacy. Evidently, the UK population did not surge overnight to account for this increase in disturbance, but hopefully the events of 2020 will be a one-off. In general, it seems very likely, although difficult to prove, that more people leads to more disturbance of wildlife. According to a 2018 analysis (Department for Transport, 2019), per-capita travel distances by people in England declined overall by about 9 per cent between 2002 and 2018. Average distances of car trips decreased by 12 per cent and walking distances declined by around 1 per cent, while cycling movements increased by 50 per cent. Apparently people have taken to cycling in contrast with otherwise consistently sedentary lifestyles. This information implies that increased disturbance of wildlife is most probably due to more people being out in the countryside, rather than to increased individual movements. Seemingly, there is a conflict of interest between advocates of more people emerging to enjoy and appreciate the countryside, and those concerned about the escalating damage to precious habitats that often results.

7 · *The Human Population and Wildlife in Britain and Western Europe*

7.1 Questions about People Numbers

The central issue of concern in this book is the extent to which the sheer numbers of humans in the UK have been an underlying factor in the problems faced by the country's wildlife. Previous chapters considered well-established specific reasons for wildlife declines and attempted to relate them, where appropriate, to increases in the British human population. It is time now to consider the facts about the numbers of people in the UK. How have these numbers changed over time, and how do trends compare with those in the world as a whole, and with those in other countries? What underlies changes in population size? There have been many previous attempts to predict demographic changes. How successful have they been in the past, and how much can we rely on them for future estimates? As the hypothesis under test here is the prediction that human population size has contributed significantly to wildlife declines in Britain, does that relationship hold up in comparison with other, broadly comparable (i.e. western, industrialised) countries?

7.2 The Global Human Population

The Story So Far

The success of *Homo sapiens* as a world coloniser is remarkable by any standards. For almost all of human history subsequent to the evolution of our species 200,000 or more years ago, and until quite recently, the global population was probably in the tens of millions. In early days, there were an estimated 300,000 or fewer people, all in Africa where the modern human story began. Genetic studies suggest that substantial increases started during the Palaeolithic period 60,000 – 80,000 years before present (BP), concurrent with a time of significant human migrations out of Africa. After the end of the Last Glacial Maximum some

20,000 years BP, human numbers escalated to perhaps 4–6 million. Then, following the widespread adoption of farming starting 10,000–12,000 years BP, the population increased to between 60 and 70 million by 6,000 years BP. Growth continued steadily thereafter to 150–300 million by 2,000 years BP, after which numbers almost stabilised over the subsequent millennium. From 2,000 years ago until around 1700 AD, the human population grew at the slow rate of 0.06 per cent per year, giving a doubling time of 1,200 years. This presumably related to the carrying capacity of the human environment in terms of food supply moderated by famines, wars and outbreaks of plague-scale diseases. Everything changed after that, commensurate with the scientific and technological advances of the Enlightenment. Most population increases during this period were in Europe, where living conditions improved and starvation became relatively rare. The growth rate consequently accelerated in the eighteenth century to 0.46 per cent per year, and further to over 0.5 per cent per year in the nineteenth century. From 1900 AD onwards, the increases in human numbers have been astonishing (Figure 7.1). Currently (as of 2020), the world population is expanding by about 83 million people annually, at 1.1 per cent per year, the equivalent of 10 cities the size of London. Human numbers increased from 1 billion in 1800 to 7.8 billion in 2020, and the doubling time at its peak, between 1950 and 1987, was just 37 years. Fortunately, the global rate of increase has recently declined and is now more linear than

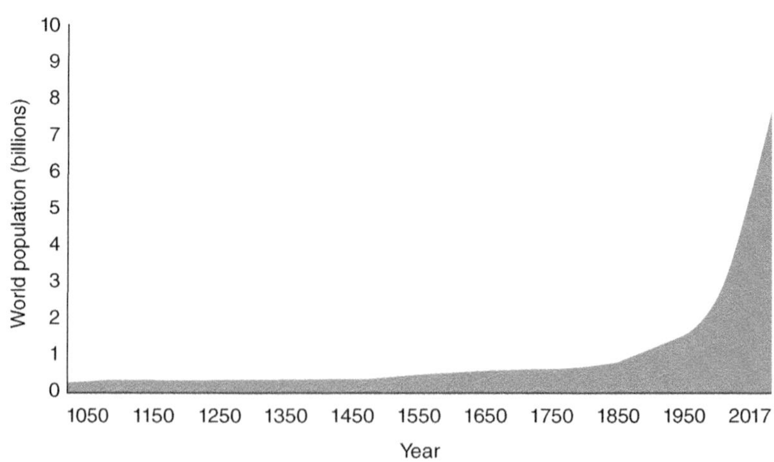

Figure 7.1. The recent history of human population increases. © Population Matters https://populationmatters.org

exponential. Nevertheless, about 15 extra people to feed are born every 10 seconds. It seems obvious that the current trend is unsustainable and will eventually be curtailed either by design or by some form of natural imposition. The latter option is unlikely to be pleasant.

How Many Is Too Many?

The trajectory of recent human population growth inevitably begs the question as to how far can it go? There must surely come a point at which the Earth's carrying capacity for people is reached or exceeded. There have been several attempts to address this concern, but the subject is a difficult one. Much depends on what assumptions underpin the calculation. At one extreme, all the world's available land surface could be converted for intensive agriculture. In that scenario, biodiversity would crash to extremely low levels. Conversely, conservation efforts could maintain and restore habitats everywhere, with substantial areas given over to rewilding. Obviously, the total possible numbers of humans would be very different in these two situations. Then there is the question of individual consumption. Per-capita utilisation of natural resources varies hugely among countries across the world. A lot depends upon whether developing countries achieve the very reasonable ambition of achieving the standard of living enjoyed by those in the long-time industrialised nations. If that happens, the carrying capacity of Mother Earth will be less than it is at the time of writing.

Despite the difficulties indicated above, calculations about limitations on human numbers have been made. An analysis of many such efforts was summarised in a report for the United Nations (UN) in 2012 (UNEP Global Environmental Alert Service, 2012). The range of the 94 estimates considered beggared belief, from 500 million to ever more ridiculous heights. In general, economists promoted much larger numbers than did scientists, perhaps because economics has too long remained detached from serious consideration of ecological constraints (see Chapter 8). Researchers at the Stockholm Resilience Centre identified nine global processes that need to stay within defined limits to sustain current 'Holocene epoch' support systems on Earth: climate change, ocean acidification, stratospheric ozone depletion, the nitrogen/phosphorus cycle, freshwater use, changes in land use, biodiversity loss, atmospheric aerosol loading and chemical pollution. Unfortunately, it looks as if three of these safe boundaries have already been passed, notably climate change, rate of biodiversity loss and changes to the global nitrogen cycle. Various studies have estimated that humans now appropriate between 24 and 32 per cent of global net primary production (NPP) for our own use. That means less NPP in the form of grass and other vegetation is available at the bottom of the food web for anything else, which in turn causes changes to the composition of the atmosphere, the level of biodiversity and the provision of important ecosystem services. Some countries with little NPP to exploit (e.g. Saudi Arabia) and other areas with many people to do the exploiting (e.g. India) have local deficits of 200–400 per cent of the local NPP. Presumably, such areas of ongoing deficit will increasingly rely on importing NPP in the form of food, fibre and materials from areas that are not in deficit.

A popular concept in considerations of limits to human population growth is the ecological footprint. Although in some ways similar to carrying capacity, instead of asking how many people could be supported on the planet in future, the ecological footprint considers only present and past years. It frames the question along the lines of how many Earths were needed to support all of the people that lived there in a given year, with that year's standard of living, biological production and technology. This leads to the idea of ecological overshoot. At present, humans use the equivalent of 1.5 planets to provide resources and absorb our waste. This means it now takes the Earth 1 year and 6 months to regenerate what we use in a year. This cannot go on. While we might reduce our per-capita footprint, the collective impact of those footprints will always be multiplied by the global population. This makes population an issue impossible to ignore. The main cluster of sustainable human population estimates falls between 8 and 16 billion people. Global numbers are fast approaching the low end of that range, and are expected to approach 10 billion by the end of the twenty-first century. Many people alive today will still be here when the planet is supporting, or not supporting, 9 billion people.

What the Future May Hold

According to L. P. Hartley, 'The past is a foreign country.' How much more so, then, is the future? History is replete with predictions of future trends in human population size, many of which have failed miserably. Even the most recent are usually accompanied with confidence limits so wide as to be of limited value. The variables likely to impact human numbers are multiple, and arguably beyond the abilities of any computer, however powerful, to engage with convincingly. Central to the problem is the variation already present in population trends across the world, which is huge and likely to remain so for many years to come. These in turn are underpinned by differences in mortality rates and birth rates, the latter increasingly influenced by fertility and contraceptive use. Life expectancy has increased everywhere, particularly since the mid-twentieth century. The improvements have been geographically uneven, however, with average extensions in Asia twice as long as those in Africa. Understanding trends in fertility rates is particularly important because these too have been very variable (Lutz & Qiang, 2002). Human numbers in the most developed countries have grown relatively slowly recently, on average at about 0.3 per cent per year, whereas in the less developed countries the rate has been around 1.6 per cent per year. Reductions in the rates of population increase since the late twentieth century have followed a so-called demographic transition, in which rapidly increasing populations with high mortality and high birth rates transform to ones with lower values in both rates. Usually mortality declines first, followed by a lag of up to a century before birth rates follow suit. This establishes a new equilibrium between birth and death rates, which does not directly cause population decline. That only happens if birth rates continue to fall. In 2020, fertility decline was well underway everywhere except in sub-Saharan Africa, and even there indications of a drop in fertility were apparent. Female emancipation, especially improved education facilities leading to widespread adoption of contraception, has been the principal cause of decreasing fertility in these less developed countries. The most important factors underpinning these changes seem to be population density, female literacy and higher standards of living, all of which correlate with reduced birth rates. Dwelling in urban areas, usually with higher incomes than prevail in rural districts, combined with school and college provision, seems to be the best recipe for containing human population growth.

Given the complexity of predictions on this important subject, what has emerged? How many people will be trying to survive on Earth over the next century and beyond? Long-term global population projections now all follow a similar methodology. Existing populations for countries or regions are grouped into cohorts defined by age and sex, and projections update numbers of each cohort annually according to assumptions about three components of population change: fertility, mortality and migration. Among these factors, fertility is the most important. Figure 7.2 shows the outcome of predictions based on this approach by the UN, for which the medium estimate is given together with confidence

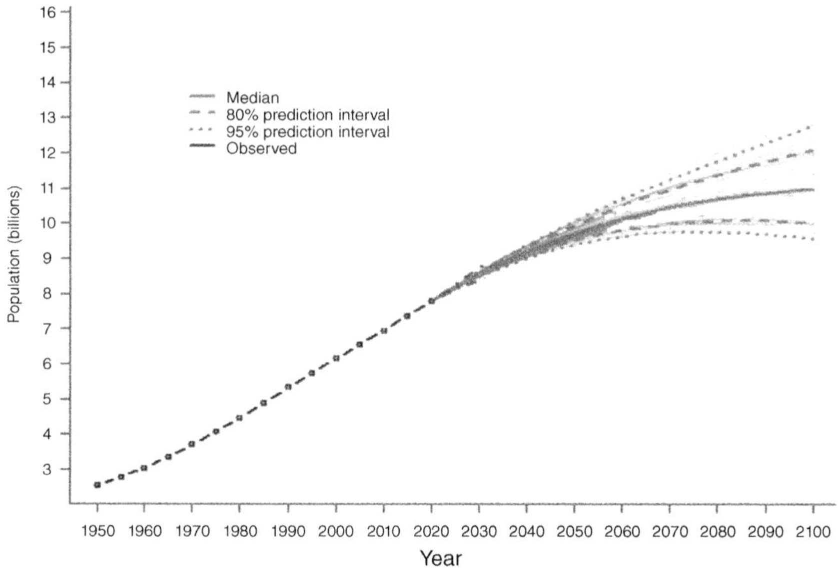

Figure 7.2. Projections of future total world human population trends. From World Population Prospects 2019, http://population.un.org/wpp. © 2022 United Nations. Reprinted with the permission of the United Nations (A black and white version of this figure will appear in some formats. For the colour version, please refer to the plate section.)

indications of projection ranges. The median estimate of future population size peaks in 2070 at 9.0 billion people, after which it slowly decreases. By 2100, the median value is 8.4 billion people with an 80 per cent likelihood of being between 5.6 and 12.1 billion. There is an 85 per cent chance that the population will stop growing before the end of the century, and by that time, there is also a 15 per cent chance that the world's population will be lower than it was in 2000. The critical determinant of these expectations is the speed of fertility decline in places where this is still high. While fertility is expected to continue declining in Eastern Europe, the populations of Africa are likely to double. In America and Western Europe, future changes are likely to depend on migration as well as fertility and mortality. And an extra factor influencing fertility rates has been identified. Religiosity, defined as the proportion of people stating in questionnaires that religion is an important aspect of their life, was positively associated with family size (Gotmark & Andersson, 2020). Mean fertility between 1970 and 1975 for the most secular countries was 2.8 children, for ones of moderate religiosity 3.3 and for the most religious was 5.4. The corresponding values in studies between 2000 and 2005 were reduced in all categories but still consistently different from one another at 1.8, 1.7 and 2.8, respectively. This is a trend in the right direction, but however grandiose computer investigations appear, there is still great uncertainty about what will actually happen.

7.3 Humans in the UK

Population Trends in Britain

Population trends in every country contribute to the global picture, but all have particular individual characteristics. Current levels of development as well as patterns of immigration and emigration vary greatly among different countries, and these factors exert influences not always directly relevant to an integrated global view. The UK is one of the most developed countries in the world and has recently experienced substantial levels of net immigration. It also has an impressive data set on human population size on account of national censuses carried out every 10 years since 1861, excepting only the war year of 1941. Reliable information about population size and trends in the UK is held by the Office for National Statistics (ONS; www.ons.gov.uk/), which is the main source of this chapter's contents. The patterns of population increase in the four countries of the British Isles over the past 1,000 years are shown in Figure 7.3. Ireland in this case includes the whole island. In 2019, using approximate numbers and percentages, England had over 56 million people (84 per cent of the UK total), Scotland had almost 5.5 million (8 per cent), Wales had 3.15 million (7 per cent) and Northern Ireland had 1.85 million (4.2 per cent). A striking feature of the comparison is the accelerating rise in people numbers in England compared with the other three members of the British Isles, starting at about 1800 AD around the time that the industrial revolution was getting underway. Overall, the pattern of population growth in the UK mirrors that of the world as a whole. At mid-year 2020, the total UK population was estimated by the ONS to be 67,886,011. Numbers were still increasing year on year in all four countries but at a progressively slower pace (averaging 0.6 per cent per year) than in any time since 2004. Rates may be decreasing, but the absolute numbers are still high, with an addition of 395,000 people to the UK (a bit more than the population of Coventry) between mid-2017 and mid-2018. Increases of this order have been normal for several years, and have somehow been absorbed into communities across the country. No doubt this trend contributes

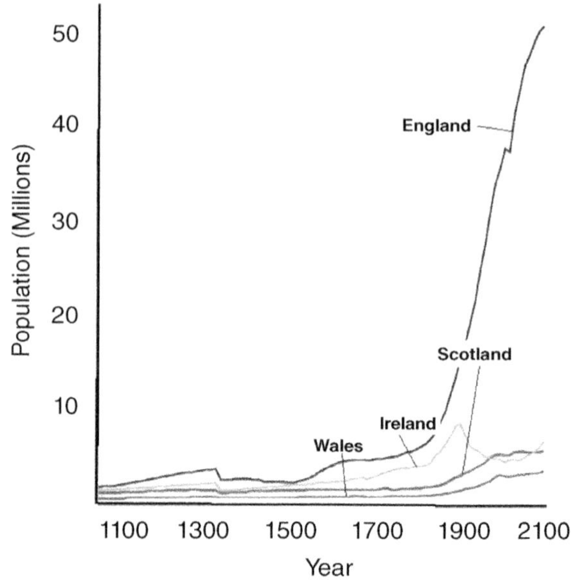

Figure 7.3. Population changes in the UK since 1000 AD. Data from ONS (www .ons.gov.uk/)

substantially towards the ongoing pressure for housing developments and expanding urban sprawl, with the downsides for wildlife described in Chapter 4.

The population of the UK is spread unevenly, with densities varying more than 100-fold from 5,700 people per km^2 across London to fewer than 50 people per km^2 in the most rural regions. In common with most developed countries, people have increasingly congregated in urban areas in recent centuries. Towns and cities are where work is easiest to find, and where social provisions such as hospitality and entertainment venues are concentrated. The downsides of city life include higher crime rates and pollution compared with rural habitats, as well as greater stress in crowded spaces. As the 2020 coronavirus pandemic showed, cities are also the places where infections spread most effectively. It is surely a remarkable achievement that a species which for most of its evolutionary history survived in small, tight-knit family groups has adapted so well to urban communities that can run into millions. In 2019, more than 82 per cent of British people lived in urban areas, leaving 18 per cent in rural spots. Moreover, the urban population has increased consistently since 1960 to almost 56 million in 2019, while over the same period the rural population shrank by 370,000 to about 11 million. Urban area definition is not straightforward, and population estimates for specific towns and cities can vary considerably according to the precise area included in the calculation. That for London, for example, was cited variously as between 7,556,900 and 10,979,000 in 2020, a difference of more than 3.4 million people. Nevertheless, the 10 most populous urban areas in the UK are more or less consistently considered to be London, Manchester, Birmingham/Wolverhampton, Leeds/Bradford, Glasgow, Southampton/Portsmouth, Liverpool, Newcastle, Nottingham, Sheffield and Bristol. Altogether more than 27 million people, about a third of everyone in the country, reside in or around these 10 places.

Causes of Population Changes in the UK

The UK underwent a demographic transition over the 200 years between 1740 and 1940. The population increased from 1760 to 1880 as death rates fell due to better sanitation and nutrition, while birth rates remained high. This was followed by a period of slower growth up to about 1940 as death rates continued to drop thanks to medical advances, but birth rates declined rapidly after 1880 due to increased use of birth control and to expanding industrialisation that generated greater prosperity. My family followed this pattern: my paternal grandfather born in 1887 was one of 13 siblings, my mother born in 1923 was one of six, and I am one of one. Since the Second World War, the UK has sustained relatively low birth and death rates compared with what came before. Nevertheless, as shown in Figure 7.3 the UK population entered into an almost exponential growth rate by the start of the twentieth century. How did this happen? For a country open to movement of people both in and out, its population size is determined by the interplay of three factors: death rate, birth rate (combined, as so-called 'natural changes') and migration rate.

Relative death rates (per number of people) in England and Wales declined continuously during the first two decades of the twenty-first century, although the decline became less acute after 2014. There was a consistent sex difference. In 2001, there were 1,500 male deaths per 100,000, but only about 1,000 female deaths per 100,000. By 2019, these figures were down to about 1,100 and 800, respectively. However, based on absolute mortalities, the changes were less marked, sex differences were reversed and the trends were not continuous. In 2001, just over 277,000 men and 293,000 women died. In 2011, a recent nadir, the corresponding numbers were 234,000 and 249,000, but by 2019, deaths increased to around 265,000 for both sexes. Differences in the two types of estimate simply reflect the greater average longevity of women. Mean age of death for men in 2018 was 79.6 years, up from 75.9 in 2001. Corresponding ages for women were 82.9 years, up from 80.5. This extra longevity has contributed to the overall increase in the UK population size. Between 1975 and 2015, the proportion of the UK population older than 65 increased from 14.1 to 17.8 per cent, and this trend is expected to continue. The major causes of mortality in 2018 were heart disease for men and dementia for women.

Birth rates have varied recently from totals of 679,000 in 2000 to 777,000 in 2015, with a peak of 808,000 in 2011. However, the long-term trend for birth rates in Britain has been consistently downwards since the early 1960s, when there were 18.3 new arrivals per 1,000 people, to only 11.4 in 2020. The total fertility rate in England and Wales has been below the theoretical replacement rate of 2.08 since 1973, and in 2019 it fell to 1.65 children per woman. Among its less serious offerings, the ONS reveals changes in the popularity of Christian names over several decades. Mine never made the top 10, peaking at 33 in 1954 – better than my wife's (Margaret, at 39 in 1964) but not as high as those bestowed on our offspring (Andrew, three in 1964, and Caroline, 15 in the same year).

Despite their declining trend, birth rates have consistently outpaced death rates in the UK such that in 2018, 'natural change' births exceeded deaths by just over 115,000. However, natural change has been nothing like sufficient to explain the UK's burgeoning population growth in recent times. ONS data collated by the Migration Observatory at Oxford University show that net migration has become an increasing driver of population change since the early 1990s. Immigration has been running at about twice the rate of emigration since the 1990s and was a bigger driver than natural change in every year from 1998 onwards, excepting 2012 and 2013. In 2018, the UK population increased altogether by over 390,000 of which natural change accounted for 115,000 (about 30 per cent of the total) and net immigration 273,000 (about 70 per cent of the total). There is also the issue of illegal immigration, which by definition is almost impossible to quantify. Home Office personnel have reckoned there are probably at least 1 million people in the UK who arrived without legal standing. The numbers of people granted asylum increased from around 7,000 in 2010 to 20,000 in 2020.

Immigration has been a politically sensitive issue in the UK, and elsewhere, for many years. Caution is needed when consulting alternative estimates of migration rates. Migration Watch (www.migrationwatchuk.org), for example, describes itself as an independent non-political think tank and claims that, from 2001 to 2016, 82 per cent of the UK population increase was 'linked' to migration – including both net immigration (roughly 50 per cent of the overall change, after accounting for immigrant deaths), and births to immigrant parents (roughly 35 per cent of the total, after accounting for deaths of people with immigrant parents). Migration Watch also puts a figure on illegal immigration, claiming the rate has run at about 70,000 per year since the mid-2010s. Independent Migration Watch may be, but it clearly has an agenda directed at reducing the numbers of immigrants coming to live permanently in Britain. Policy decisions about the control of immigration have received even more attention than previously since the UK left the EU, and are unlikely ever to attain universal approval.

Immigration into Mainland Europe

About 2.4 million immigrants entered the EU from non-member countries during 2018, an increase from 1.5 million in 2013. Across Europe as a whole, in 2018 the UK accepted the third highest number of total, not net, immigrants (604,000) out of the continent's 32 countries that provided data. Some did not, notably Ukraine, Serbia, Albania, Bosnia-Herzegovina, Montenegro, Kosovo, Moldova, Russia and Belarus. Only Spain and Germany took in more people than the UK. The traffic was not all one way, although most (22 of the 27) EU Member States registered more immigration than emigration in 2018. Only in Bulgaria, Croatia, Latvia, Lithuania and Romania did the number of emigrants exceed the number of immigrants. The figures look somewhat different if immigrant numbers are compared per size of the receiving population, with the UK on this basis coming 19th out of 32. The smallest countries tended to have the highest migrant to resident ratios, with Malta, Luxembourg, Iceland and Cyprus topping the list. Once in Europe, the fate of immigrants is complex as many move among the constituent countries. Asylum seekers entering the EU are supposed to register where they first enter, but many do not and have a target country in mind that they attempt to reach. This ambition can be driven by the presence of relatives already in that country, or perhaps by a national language with which they are already familiar. Those attempting to cross the English Channel by boat from France evidently do not seek asylum where they enter Europe, but choose instead to undertake a perilous sea crossing where numbers perish every year. Some 5,000 people entered the UK this way in 2020. At Europe's southern border, the situation is even more desperate. Nearly 124,000 immigrants crossed the Mediterranean by boat from Africa in 2019, and increasing numbers of people every year try to cross this dangerous piece of water, travelling mostly from Libya to Italy or Spain. More than 1,300 people drowned in 2019 attempting this journey. Those immigrant refugees surviving the trip face increasing hostility in the countries they reach, some of which feel overwhelmed by the numbers involved. Immigration is a global issue likely to increase in future, especially if the already damaging effects of climate change get worse in Africa and Asia.

Predictions of Population Trends in the UK

As with the global predictions of population trends, equivalent estimates have been made for what is likely to happen in the UK. The ONS projects an increase of 3 million people, 4.5 per cent of the 2018 population, to 69.4 million by 2028 for the country as a whole. However, sharp differences are anticipated among the UK's constituent countries, with the English population predicted to expand by 5 per cent, that in Northern Ireland by 3.7 per cent, that in Wales by 2.7 per cent and that in Scotland by 1.8 per cent. More than 70 per cent of these increases will be due to net immigration. Looking further forward, there is an expected UK population of 70 million by 2031 and of 72.4 million by 2043. Across this longer period of 25 years, England will host 6 million additions, by far the largest number of new residents within the union. By contrast, Scotland and Wales will each gain an extra 200,000, and Northern Ireland just 100,000. As the years go by, death and birth rates are likely to balance out such that an even higher proportion of the later population increases, perhaps 83 per cent, will be due to immigration.

There are other recent estimates of UK population growth, based on different algorithms, that predict larger numbers than those calculated by the ONS. Thus, the EU Statistical Office (Eurostat; https://ec.europa.eu/eurostat) projects 75.3 million, 3.5 million more than the ONS estimate, by 2040. On Eurostat's reckoning, the UK's projected growth of 14 per cent between 2018 and 2040 is much higher than the EU average and is highest among the EU's four largest nations (including, before Brexit, the UK). France's population is expected to grow by

6 per cent and Germany's by 1 per cent, while Italy's is projected to decline by 5 per cent. The UN has also had a go at forecasting what will happen in Britain, proposing a population of 73 million in 2043.

Evidently different analytical methods have yielded results that in some cases vary quite widely from one another. It is also instructive to look at how earlier attempts to forecast population changes have fared. The ONS predictions of 2016 were already out by 2018, after calculating a population 30,000 larger than actually materialised in that year. The first well-documented attempts to predict trends in the British population were made in the 1950s, and others followed at various times thereafter. Inevitably, there were considerable variations in the accuracy of these efforts, and those made during the latter half of the twentieth century were assessed by Shaw (2007). Projections require estimations of current population size, mostly accurate in the UK due to the decadal census returns; concurrent birth and death rates, also well documented; and migration rates, the least certain of the available data. Projections tended to assume that birth, death and migration rates would remain broadly similar or follow already evident trends. These expectations have not always been met. Birth rates increased unexpectedly during a 'baby boom' in the 1960s, after which they returned towards pre-boom levels. Death rates fell consistently, defying a common notion that they would stabilise and then fail to improve beyond a point regularly thought to be close at hand. Migration rates proved the most flirtatious. Until the early 1980s, more people emigrated from the UK than entered the country. After that time, the trends reversed, and immigration became dominant. This trend accelerated dramatically in the early 2000s, and has continued in the same vein ever since.

How, then, did previous predictions fare? From a baseline of 50 million in 1950, the 1955 projection for the following 40 years anticipated a population of England and Wales of 53 million in 1995, 5 million fewer than the actual number. This is a projection I remember noticing at the time because I was already concerned about wildlife issues, especially the disappearance of ponds under housing estates. It turned out to be unduly optimistic. By 1965, the baby boom had changed the birth rate baseline and predictions from that year, now for all of the UK, assumed it would stay at about three children per woman, compared with about 2.3 in 1955. This produced the highest proposed increase in population of any prediction, anticipating 75 million people in 2000 and a steady addition of 1 million people per year thereafter. The actual number in 2000 was 59 million, a 16 million deficit relative to the projection. But the baby boom was a blip, and by the 1977 projection date, fertility rates had dropped back down to a record low of 1.7 children per woman. Taking this change into account, as well as net emigration, this model predicted that there would be just under 58 million British people in 2011. This was the lowest number of any prediction for that year, in which the actual number was 63 million. Subsequent predictions have been too recent to ascribe changes over periods comparable in length to earlier ones. Those in the 1980s and 1990s assumed continued low fertility, reduced death rates and low net immigration leading to population stability in the 2020s. Once again, however, circumstances changed, and by the time of a 2004 projection, higher levels of immigration forecast an increasing population until at least the 2070s.

In retrospect, it is possible to quantify the error rates in UK population predictions. In 2005, the actual population was 60.2 million. Predictions for that year since 1970 ranged from a maximum of 64.3 million in 1971 (6.8 per cent too high) to a minimum of 57.5 million in 1977 (4.4 per cent too low). Aside from that of 1971, all projections made in the 1970s and 1980s underestimated the total population in 2005. Since 1991, there have been errors in both directions. There is little to suggest that the accuracy of predictions has improved over the last 50 years, and given the unforeseeable changes, particularly occasional surges in birth rates and immigration, this is hardly surprising. The unpredictable pattern limits any faith that can be placed on the accuracy of projections published by the ONS and others, which are themselves variable according to the algorithm employed, as described in earlier sections. Of course, it is important to attempt an understanding of how the UK population will change in future, but such a moveable feast confounds confidence in what the numbers will actually be 10, 20 or more years down the line.

7.4 Human Populations and Wildlife beyond the UK

Making Sensible Comparisons

A well-known example of human population increases impacting on wildlife is the ongoing encroachment of people on the wild savannahs of East Africa. The Greater Serengeti ecosystem, comprising a national park, adjacent reserves and surrounding controlled areas, is a prime example of escalating human–wildlife conflict (Estes et al., 2012). Pressures working against the conservation of such jewels in the crown in the world's wild places follow from ever more humans expanding agriculture as their numbers increase, resulting in the hunting of bush meat and the poisoning of apex predators such as lions and hyenas. But African ecology is hugely different from that of northern temperate countries such as the UK and mainland Europe. Human population densities are much higher in the north than in tropical Africa. There are no vast plains with enormous herds of herbivores and attendant predators, but instead a much more organised countryside dominated by farming activities and forestry, together with urban sprawls. It was not always thus. In the Pleistocene period, vast numbers of mammoths, bison, deer and other grazers swarmed across the 'mammoth steppe' of Eurasia, with an accompanying suite of predators. It has become increasingly certain that their demise was triggered by the arrival of modern humans with their novel hunting techniques. What is happening in Africa today mirrors events in the late Pleistocene in Eurasia, although hopefully it will not be followed by the species extinctions characteristic of that time. It follows that the most appropriate comparisons of recent wildlife declines in the UK are with countries as similar as possible with respect to human numbers and landscape features. That means Western Europe.

Wildlife Declines in Western Europe

No European country has remained untouched by environmental degradation, but the extent of it varies considerably across the continent. As in the UK, urbanisation and intensive agriculture have often been cited as the main causes of damage to wildlife, especially where human populations are high and arable farming is widespread. There are difficulties in trying to make quantitative comparisons among countries because very often the types of measurements made to detect trends in species abundance differ, as do the time frames over which records have been kept. For some taxonomic groups, there is little or no transcontinental information. The UK stands out as one of the best documented countries in the world for wildlife studies, some extending over many decades, but most investigations elsewhere are limited to relatively few prominent groups of well-studied organisms. Information for this book was sought in relation to wildlife declines in Austria, Belgium, Denmark, France, Germany, Ireland, Italy, the Netherlands, Portugal, Spain and Sweden. East European countries were excluded because their histories of development, and particularly of agricultural practices, have until relatively recently differed substantially from those in the west of the continent. All 11 of the countries selected are members of the EU, which produced a

State of Nature report (European Environment Agency, 2020) painting a moderately bleak picture of habitat and wildlife trends between 2013 and 2018 in Europe. Its habitat assessments concluded that 81 per cent were in poor or bad condition, and more than 36 per cent were deteriorating. Dunes, bogs, mires and fens topped the list of habitats failing to maintain good status. Heathlands declined enormously during the twentieth century, together with specialised plants and animals characteristic of this oligotrophic (base-poor) habitat. The report also concluded that less than half of the continent's birds had a good population status and about 30 per cent of them were declining. Species assessments were also attempted for other species groups. The results indicated poor or bad conservation status for about 55 per cent of vascular plants, more than 70 per cent of non-vascular plants, at least 65 per cent of arthropods and molluscs, 55 per cent of other invertebrates, 80 per cent of fishes, 60 per cent of amphibians, about 50 per cent of reptiles and 55 per cent of mammals. Overall, more than 60 per cent of these assessments judged their subjects to be in poor or bad status. As for trends, 35 per cent of the species assessments with an unfavourable or unknown status indicated continuing declines. It is important to put these gloomy figures in a broader perspective. Among all these groups there were species improving their lot. Top of this list were 36 per cent of vascular plants and 40 per cent of reptiles. There were also quite a lot of species with unknown status, often 10–20 per cent of the total but as many as 30 per cent of 'other invertebrates'. Nevertheless, as in the UK an overall pattern with a large number of plants and animals undergoing declines seems to be universal.

Among the groups listed by the EU State of Nature report, four are of particular interest. Firstly, vascular plants are important as primary producers. Their demise, mostly due to habitat loss, can have serious consequences for many other types of wildlife including pollinators and grazers, not to mention other animals in the food web. Secondly, insects are by far the most common and diverse group of invertebrates. They have attracted much attention since the revelation that protected areas in Germany experienced a 75 per cent reduction in flying insect biomass over 27 recent years (Hallmann et al., 2017). To entomologists in Britain, this will have come as little surprise, as the Rothamsted Research station has been documenting moth declines in that country for several decades. Again habitat loss has been a primary cause, supplemented by the industrial-scale application of lethal pesticides. Insects are vital components of thriving food webs, and also as pollinators of wildflowers and many commercial crops. Their decline is widely considered to underpin the downward population trends of many farmland birds that feed on them. Thirdly, amphibians have decreased and too often become extinct across the world at a higher rate than any other vertebrate group, mostly consequent upon habitat destruction and the emergence of novel, highly pathogenic diseases. Finally, birds are a biodiverse group of animals that have been relatively well monitored, with the result that there is more trend information for these animals than any other. How have these members of Western Europe's wildlife heritage fared in various individual countries?

Wildlife Decline in Specific Countries

Even within the relatively limited area of Western Europe, countries vary enormously in size, habitats and climate. Human population density is also highly divergent, with differences across the 12 countries in this comparison of more than an order of magnitude. At the two extremes, Sweden has an average of just 23 people per km^2 while the Netherlands has 421. At 280, the UK is well above the average of 181, and England is highest of all with 430 people per km^2. This broad range of human numbers across the continent provides a good opportunity to test the hypothesis that wildlife declines correlate with human population density. The comparison covers countries with broadly similar economies and development histories but may potentially be confounded by differences in landscape and climate. Those in the north have rather different habitats and species from those in the south, as well as lower overall biodiversity. Much depends on the extent of information available from each country, and these are considered, in alphabetical order, below. In all cases, the examples of declines are just a sample of multifarious reports and are not by any means a comprehensive account of the happenings in each state.

Austria

Austria experienced a general decrease of vascular plants at 95 sites of species-rich grassland, from an average of 43 species in 1990–1992 to 31 in 2011. Much of this country is blessed with impressive mountainous scenery, and although agricultural intensification is ongoing in the uplands, glorious floral displays in the alpine meadows are, happily, still widespread. Insect declines have only recently attracted international attention, and no long-term studies are yet available for Austria.

Among the amphibian community, common frog declines have been noticed in the Salzburg district, whereas common toad populations seem to be stable. Curiously, frog declines have been most marked in alpine areas rather than in the more intensively farmed lowlands. However, farmland birds have certainly been in trouble, with estimated recent decreases of 30 per cent.

Belgium

Belgium is home to an impressive variety of habitats including coastal dunes, inland heaths and mountain woodlands. As in so many countries, much of the lowlands have become a victim of industrialised agriculture and wildlife has suffered in consequence. Widespread eutrophication due to fertiliser applications remains a major problem. Heathland in particular has come to grief, with substantial losses of vascular plants associated with nutrient-poor soils. The Campine region formerly consisted of a vast heath landscape, stretching from the River Scheldt in the west to the River Maas in the east and the north. Now only 15 per cent of this heath still exists, stretching across the Belgian and Dutch border. There is not much information on the fate of most insects, but a monitoring programme showed that 19 of the 64 butterfly species previously indigenous to the Flanders region have been extirpated there. Some 27 per cent of all Belgian vertebrates including 28 per cent of mammals, 57 per cent of reptiles, 44 per cent of amphibians, 30 per cent of birds and 18 per cent of fish were considered threatened in 2017. Since 1990, 60 per cent of Belgium birds have declined.

Denmark

Denmark is mostly a peninsula flanked by the North Sea and connections to the Baltic. Its generally flat terrain includes dramatic coastal dunes and formerly extensive heaths, bogs and marshes. However, many heaths have gone under the plough, and modern agriculture dominates the inland scenery. The country's heath society was set up in 1867, not for conservation but to reclaim the bogs and moors for farming. It proved all too successful, and relatively little of this wild habitat now remains. One victim of heathland loss was the smooth snake, and Denmark has the unenviable reputation of being the first country to lose this secretive reptile. Another impact of modern agriculture has been the decline of 60 per cent of arable weeds since 1990. Invertebrates are also having a hard time. Based on counts of flying insects killed on car windscreens as a measure of insect abundance, between 1997 and 2017 there were reductions of 80 and 97 per cent at two transects of 1.2 and 25 km, respectively. Reports suggest that all of Denmark's amphibians declined as agricultural intensification took hold after the Second World War, as have perhaps 50 per cent of its farmland birds.

France

France is the largest country in the EU in terms of surface area, and is more than twice the size of the UK. It extends from the cool English Channel coast south to the balmy shores of the Mediterranean, encompassing many types of habitats along the way. Coastal dunes, inland heaths and marshes, forests, smallholder farms and high mountains support great levels of biodiversity. As elsewhere, though, there are also extensive areas of relatively uninteresting (from the wildlife perspective) intensive agriculture, and France has been one of the highest consumers of pesticides. No doubt this and diminished habitat has contributed to wildlife declines in the country, some of which have been substantial. At forestry plots in north-west France, species richness of vascular plants was reduced in 2009 relative to earlier surveys in 1976, a period during which bryophyte diversity increased. Systematic measurements of insect numbers have not been documented, but surveys have shown that a majority of people in France are concerned about pollinator losses. The status of amphibians has been reviewed in detail. Of the 34 native species, 15 were listed as under some degree of threat in 2013. Two of these, Lanza's alpine salamander (*Salamandra lanzai*) and the moor frog (*Rana arvalis*) were considered to be seriously endangered. Between 2008 and 2018, many birds declined in all the habitats monitored but most markedly in agricultural areas. Other studies specifically showed a 33 per cent decline in the populations of common farmland bird species after 2001. Numbers of meadow pipits (*Anthus pratensis*) fell by 68 per cent in 17 years, while linnets declined by 27 per cent over the same period. Populations of skylarks and grey partridge are collapsing, with declines of 50 and 90 per cent, respectively, over 25 years.

Germany

Germany is the most populous state in the EU. It has the same broad range of habitats that occur in Belgium and France, including coastal dunes and marshes, heathlands, forests and mountains, as well as extensive areas of agriculture. A familiar pattern emerges. Since the 1960s, more than 70 per cent of the vascular plant species investigated in a wide-ranging study showed significant nationwide declines, apparently a result of several different but unidentified causes. Undoubtedly, though, habitat loss and degradation have contributed to the problem. As in much of north-west Europe, heathlands were formerly widespread in northern Germany but have been widely encroached on by other forms of land use, leaving mostly just isolated fragments. Many of the rare plants and animals that thrive on heathland have suffered in consequence. It was in Germany that the astonishing declines in flying insects were quantified, as mentioned earlier. A big surprise was that this problem occurred on nominally protected conservation areas. However, these places were usually surrounded by farmland. Adjacent agricultural intensification, including pesticide usage, year-round tillage, increased use of fertilisers and other associated activities were considered possible causes. On top of that, perhaps half of Germany's amphibians have lost ground in the post-war decades, with fire-bellied toads (*Bombina bombina*) of special concern. Birds, too, have taken a tumble. Twenty-three of the 64 commonest German birds have fallen in numbers over the last 15 years. House sparrows, house martins and lapwings continue to lose ground.

Ireland

Ireland remains a predominantly rural country, a European outpost in the Atlantic. Its isolation has resulted in relatively low biodiversity, but there is nevertheless a superbly rich landscape with coastal dunes, extensive bogs, impressive mountains and the unique limestone pavement of the Burren. Unfortunately, Ireland's extensive and precious bog system has suffered widespread damage from the expansion of agriculture and forestry. Raised bogs are practically extinct, and of the previously vast blanket bogs, only 28 per cent are now deemed worthy of conservation. Wildlife in Ireland has experienced a chequered history. Wolves survived there probably longer than in the rest of the British Isles. The final definite sighting was in 1786, at least a century later than the last reliable record from Britain. By contrast, the golden eagle was extirpated from Ireland around 1912 but has maintained a stronghold in Scotland. Ireland's rural idyll has not, unfortunately, prevented more recent difficulties for wildlife from arising. Of 1,047 vascular plants investigated, 8.8 per cent were judged to be at risk in 2016 according to IUCN Red List criteria. There is precious little historical information about insect numbers in Ireland. However, over 30 butterflies are under serious threat, and one species, the mountain ringlet (*Erebia epiphron*), has become extinct. There are only three amphibians in Ireland, one of which has always been rare. The natterjack toad disappeared from many localities in the twentieth century, leaving a fragmented pattern of isolated populations. Irish birds have had hard times recently. Of 12 species of wader nesting in Ireland, 11 are in serious trouble. Only around 150 nesting pairs of curlew remain in Ireland, down from 150,000 pairs in the 1970s. About 25 per cent of farmland birds are also declining. For all of these groups, intensification of agriculture including improved drainage has been the main driver of wildlife losses.

Italy

Italy is a country of hills and mountains with lakes, forests and river valley wetlands. Its southerly location saved it from direct effects of the Pleistocene glaciations, leaving it with a highly diverse flora and fauna. Bathed in tepid Mediterranean seas, Italy's climate is warmer than that in northern Europe, although the high mountains are a lot colder than the coastal lowlands. Intensive agriculture is mostly in the north of the country, where its impact has been along the same lines as elsewhere in Europe. Of 2,430 native vascular plants registered, 2.2 per cent of those assessed were thought to be nationally or regionally extinct and 24.3 per cent were considered to be at risk in 2021, based on IUCN Red List criteria. As with most countries, there is little information about trends in insect diversity or abundance, but there have been occasional studies of specific groups. Thus, an investigation of roller dung beetles revealed a 31 per cent reduction in both the abundance and distribution of 9 out of 11 species during the twentieth century, starting in the 1960s. Studies of amphibians showed that 11 per cent of 203 breeding sites present in 1979, sampled across the entire country, had been destroyed by 2008. According to IUCN Red List criteria, 22 per cent of Italy's amphibian species were threatened in a 2013 assessment. Italians are notorious for hunting birds, especially migrants that move from Africa to northern Europe in spring and back again in autumn following a summer breeding season. The casualty numbers are appallingly high, at more than 5 million birds killed annually, but it remains unclear as to what contribution shooting makes to severe migrant population declines. Habitat loss in their countries of destination is surely another, quite probably major, cause of the issues faced by these wanderers. Nevertheless, the carnage has continued year on year, despite EU legislation designed to control it but which is scarcely enforced. Hunting is certainly not the only cause of bird losses in Italy, and agriculture has again played its part. In Lombardy, 7 of 51 bird species declined whereas 10 improved their lot. Big losers were red-backed shrikes (*Lanius collurio*) and skylarks, both of which decreased by more than 70 per cent between 1992 and 2007.

The Netherlands

The Netherlands is a small, densely populated country, much of which lies close to or below sea level. There are impressive coastal dunes, low-lying marshes and extensive heaths, as well as the inevitable intensive agriculture. The Netherlands has been blessed with a long tradition of natural history interests, and many species are well recorded. Arable vascular plants have declined by at least 35 per cent since 1900. An important change during the twentieth century was a strong reduction in oligotrophic and a marked increase in eutrophic plant species, presumably consequent on nitrogen enrichment via fertilisers, with habitat fragmentation also identified as an important cause of plant declines. Measurements over the recent decades have indicated serious insect declines, with 54 per cent of ground beetles, half of 47 butterfly species measured and 62 per cent of caddis flies all on the wane in various parts of the Netherlands. Nine out of 16 amphibians were red-listed by the IUCN in 2007. Species decreasing in numbers include natterjack toads, fire-bellied toads and both smooth newts and palmate newts (*Lissotriton helveticus*). The fire salamander (*Salamandra salamandra*) has undergone a decline almost to extinction in the Netherlands following outbreaks of the accidentally introduced pathogenic fungus *Batrachochytrium salamandrivorans*. Between 1990 and 2014, numbers of meadow birds declined very rapidly. Black-tailed godwits dropped by 60 per cent, lapwings by 55 per cent, oystercatchers by 70 per cent and redshanks by 33 per cent. In the period from 1990 to 2017, 13 out of 20 city-breeding bird species also declined. The crested lark (*Galerida cristata*) has completely disappeared from the country, at least as a breeding species. Overall, though, the news is not all bad. Of 195 birds investigated, 33 per cent have shown a downward trend since 2007 but 39 per cent have increased.

Portugal

Portugal, set on the Atlantic face of the Iberian Peninsula, is a land of coastal dunes, marshes, high mountains, grasslands and woodlands that include rich cork oak forests. Far enough south to escape the ravages of

Pleistocene glaciations, the country has a rich biodiversity heritage. But like elsewhere in Europe, there have been significant wildlife declines. A national list of all the species either in grave danger or on the brink of extinction includes nearly half of the country's wildlife including the enigmatic Iberian lynx (*Lynx pardinus*). This situation is a result of years of intensive development and the introduction of foreign species, particularly exotic eucalyptus trees, which were planted extensively in the 1960s. The country has also been plagued by wildfires, damaging large swathes of woodland habitats. In 2018, over 17 per cent of Portugal's vascular plants were judged to be threatened. So far there has been little monitoring of insects in Portugal, but one estimate reckoned that half of the nation's cicadas face 'various threats'. In some parts of Portugal, local amphibian declines have followed the appearance of North American signal crayfish, voracious predators, especially of tadpoles. Some local impacts of chytrid infection, especially killing midwife toads (*Alytes obstetricans*), have also been reported. Changing agricultural practices have had significant, sometimes serious, impacts on bird populations. They probably account for the drastic 49 per cent decline of little bustards (*Tetrax tetrax*) between 2006 and 2018. Development of local roosting sites was associated with major decreases of some abundant waders in the Tagus estuary, notably dunlins, grey plovers (*Pluvialis squatarola*) and redshanks over 30 years.

Spain

Spain, a large country encompassing most of Iberia, is a naturalist's paradise. Untroubled by Ice Age glaciations, the mountains, grassy plains, upland forests, cork oak forests, coastal dunes and multiple wetlands harbour a huge range of plants and animals. But yet again, this wildlife heritage has not survived unscathed. There have been extinctions of vascular plants, but in general declines across so vast and complex a country have been hard to measure. Similarly, for insects there are essentially no data documenting recent trends currently available. For the squeamish, however, there is a website listing 'dangerous creepy crawlies in Spain (insects and reptiles)' (www.benidormseriously.com/dangerous-insects-and-reptiles-in-spain). These include 'nasty' Asian hornets, 'dangerous' hairy caterpillars and black scorpions (which are neither insect nor reptile) that 'like resting in your shoes'. So beware. There are some well-documented amphibian declines, mostly local and due to pathogens such as ranavirus and chytrid fungus outbreaks.

Midwife toads and common toads (*Bufo spinosus*) have been the most frequent victims. Unlike most other taxonomic groups, there is considerable information about bird populations, and most of it signals bad news. Many species are in increasing difficulty on account of reductions in fallow land due to changing patterns of agriculture. Fallow land is cultivated but not seeded for one or more growing seasons. It includes semi-natural grasslands and pastures that will eventually be ploughed for a new crop cycle but which are one the country's richest wildlife habitats. Apparently Spain lost 64.5 million birds between 1996 and 2015, more even than the UK, which has lost around 40 million since 1980. Between 1998 and 2005, common rural birds declined by 23 per cent while common urban birds diminished by 18 per cent. Turtle doves, those welcome but rapidly diminishing summer visitors in England, declined by 37 per cent between 1996 and 2018. Large numbers of these increasingly endangered birds are still shot annually by hunters. Little bustards declined by 71 per cent, and common quails (*Coturnix coturnix*) and yellowhammers each by 50 per cent. Most of the downturns are due to the same old suspects: conversion of dry grassland and low-intensity cultivation to intensive arable agriculture, including the planting of monocultures and perennial crops, irrigation and afforestation. At least great bustard numbers seem to be stable in Spain and the plains of Extremadura remain a wonderful place to spot these impressive aviators.

Sweden

Sweden extends the full length of the Baltic Sea, from Arctic tundra in the north to the warm southern outpost of Skåne. Within the EU, only Finland has a lower human population density than Sweden, and the country boasts a plethora of lakes, mountains and vast forests. Most intensive farming occurs in the central lowlands, and in the far south lie other habitats including heaths and calcareous grasslands. Wildlife diversity in northern countries is generally lower than in warmer ones, but Sweden supports populations of particularly interesting animals including Eurasian elk (*Alces alces*), brown bears and wolves. Despite the wild nature of much of the country, plants and animals in the south have suffered declines, and some habitats have deteriorated. A combination of eutrophication from fertilisers and climate warming is held responsible for vascular plant declines, and in Skåne, arable weeds, grassland species and plants growing on uncultivated ground have decreased by about 23 per cent since 1987. Oligotrophic freshwaters and heaths now comprise the country's most rapidly declining vegetation types. Swedish entomologists developed a 'malaise' trap for recording flying insects, but a nationwide survey only started in 2003 and no trend data are yet available. The wonderful wilderness of the country has caused a few interesting problems with the scheme. One trap was destroyed by a moose bull rubbing his antlers against it, and another was eaten by a group of reindeer. Several amphibians reach their northern range edge in Sweden. Some have declined, including great crested newts, spadefoot toads (*Pelobates fuscus*) and fire-bellied toads. As usual, habitat destruction including pond loss has been a primary cause. Swedish birds also are suffering from habitat change. Between 2000 and 2018, 55,000 ha of woodland were logged in the western parts of northern Sweden and many of the country's woodland birds are decreasing, especially in lowland areas. Of 21 species associated with farmland, 71 per cent underwent significant declines between 1976 and 2001, all connected with agricultural intensification.

Comparisons of Wildlife Declines among Countries

What can be gleaned from documenting wildlife declines in this range of west European countries? Firstly, it is clear that nowhere has proved immune to deterioration of habitats and species. On the face of it, this is unsurprising as it is well known that these problems are global. However, much attention has focused on biodiversity disasters in developing countries, especially the ongoing destruction of rainforests. These trends are driven largely by poverty, often fuelled by increasing human populations. Less often are comparable issues in industrialised, affluent societies put under the microscope. It seems difficult to understand why rich economies have so often allowed wildlife declines to continue unabated. The answer may relate to the commonality of agricultural intensification as a driver of wildlife declines across Western Europe. This in turn has coincided with increasing human numbers and corresponding needs for high levels of food production.

One way of pursuing this issue is to compare quantitative variations in wildlife declines across the European countries considered in this chapter. Of the four taxonomic groups under scrutiny, two proved unsuitable for identifying such relationships. In the case of vascular plants, the available information was too fragmentary for formal analysis, and for insects, serious investigations of declines have begun too recently to provide enough data. For the other two categories, the situation is better.

Corbett (1989) collated information about amphibian declines in all European countries since the Second World War. The compilation included results from long-term monitoring moderated by expert assessments to provide species trends. As shown in Figure 7.4(a), overall amphibian declines (percentages of species present) correlated positively with human population densities in the 12 countries (those listed in this chapter, and including the UK). The same was true for bird declines (Figure 7.4b). EU member state data were used to analyse bird population trends based on a standardised methodology (European

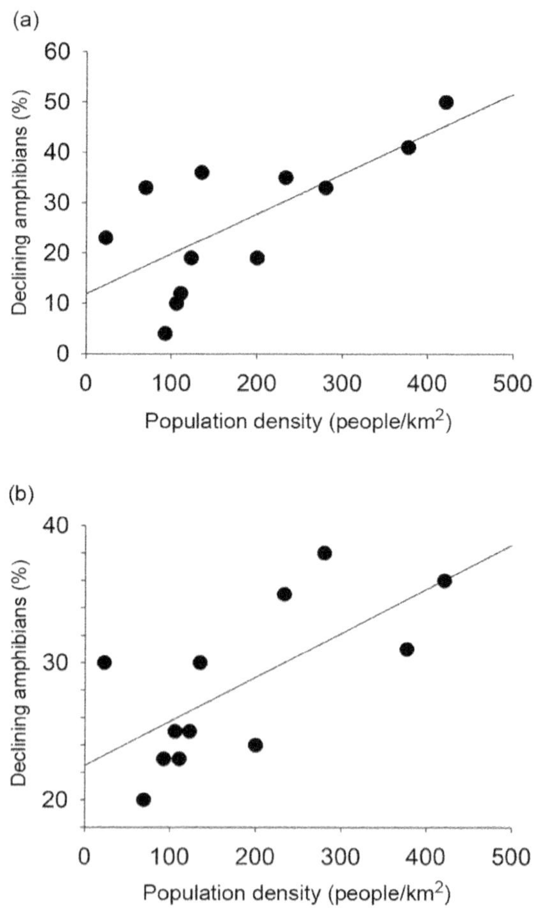

Figure 7.4. (a) Percentage of amphibian species declining. (b) Percentage of bird species declining. Both are shown as a function of human population density

Environment Agency, 2020). Human population densities are taken from Table 1.1 in Chapter 1.

In both cases, the relationships between declines and population density were statistically significant (amphibians: Spearman rank correlation (r_s) = 0.660, $P = 0.022$; birds: $r_s = 0.724$, $P = 0.010$). Correlations do not necessarily imply cause and effect, but given the information described in earlier chapters, as well as in this one, about direct causes of wildlife declines, an influence of human population size seems credible. Considering how much variation exists in other features among these countries, the relationships of Figure 7.4 are striking. The inter-country comparisons are consistent with the founding hypothesis of this book.

7.5 Overview

For most people, it is hardly news that the human population increased dramatically in recent decades and is continuing to do so. The world in general, and the UK in particular, are in population overload (Figure 7.5).

The global trend has begun to tail off, but several studies suggest that even current numbers are not sustainable in the long term and some key resources are already oversubscribed. Wildlife has become one of many

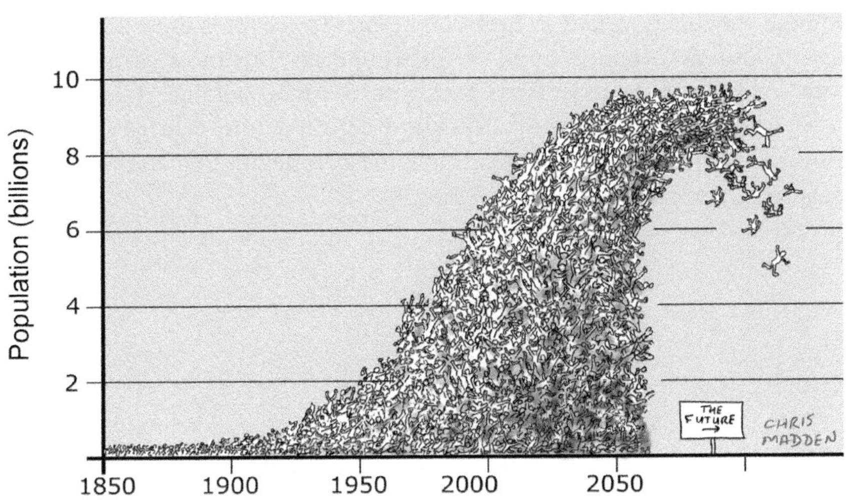

Figure 7.5. Going over the top: the global population tsunami. © Chris Madden

casualties of overpopulation, not just in the world as a whole but as much, if not more so, in affluent developed countries. The best guess, based on various computer models, is that human numbers will peak at around 9 billion by mid-twenty-first century, but there are wide error limits on this expectation. We can anticipate a lot more population-based stress in the coming years.

The pattern of population trends in the UK has broadly mirrored that of the entire planet. The UK's 2020 population of 68 million is predicted to reach more than 72 million by 2043. Here again, there are substantial uncertainties based on several different forecasting algorithms. What is clear is that of three possible engines of recent population growth in the UK, the birth rate has not been a major one. For many years, this has been mostly below replacement level. Living longer, another possible causative factor, has contributed to recent increases, but high levels of immigration have been the dominant force. This situation is predicted to continue, albeit dependent on government policy. Crowded countries like the UK, especially England, are currently set to become even more so.

Comparison with the UK's nearest neighbours reveals that wildlife declines beset all of them. It also shows that, at least for groups with sufficient data to analyse robustly, notably amphibians and birds, the extent of these declines relates significantly to the densities of people living in each country. This has implications for how wild plants and animals are likely to fare in future. The present chapter only documents recent, usually negative trends in many species, but most countries are now making increased efforts to conserve their wildlife. Chapter 10 considers these activities and discusses how their future success may be constrained by ever more humans making competing demands on high-quality green space.

8 · *Public Perceptions of Wildlife and Population Issues*

8.1 Public Opinion and Wildlife Conservation

Wildlife has many fans among the British public. A YouGov poll sponsored by the RSPB in 2020 revealed that 84 per cent of respondents wanted the government to increase the number of accessible, nature-rich areas; 80 per cent rejected the idea of less government spending on wildlife conservation; and 76 per cent reckoned that nature could contribute to economic recovery in the UK. A pity, perhaps, that this poll came 10 years after the government began devastating cuts of 50 per cent to Natural England, that country's statutory wildlife conservation agency, leaving it barely fit for purpose. In a previous poll, the overwhelming majority of Brits voted for wildlife conservation measures post-Brexit at least as strong as those in the EU. In yet another reality check, the public was asked to list the most pressing issues facing the UK in 2021. Out of 14 possibilities, the environment came out as the fourth most important, cited by 26 per cent of respondents, surpassed only by health (60 per cent), the economy (57 per cent) and leaving the EU (29 per cent). Lower down the agenda were immigration and asylum, crime, defence and security, tax, pensions, education, family life and childcare, housing, transport and welfare benefits. These results reinforce a long-standing disparity between what most politicians and the rest of us see as matters of high importance to our well-being. It seems that most of our decision makers focus on economic concerns reckoning, probably rightly, that when it comes to the ballot box it is income and job security that determines what most people will do.

Perhaps more surprisingly, in another YouGov poll, 82 per cent of people supported the reintroduction of species that have become extinct in Britain since the last Ice Age. Less remarkably, the support for reintroductions was very variable according to the species concerned. Of the many options, beavers are already with us and 62 per cent of the public (and 76 per cent of those who want to reintroduce missing species)

Figure 8.1. Beavers are back, to the delight of many people in the UK. *Source*: Arterra/Universal Images Group/ Getty Images (A black and white version of this figure will appear in some formats. For the colour version, please refer to the plate section.)

would approve of their reintroduction on a larger scale (Figure 8.1). White-tailed eagles are another reinstated success story, and most people would be happy to see more of these and of other currently rare or extinct birds. Wild boars made an impromptu comeback by escaping from farms, and now survive as several feral populations in England. About 38 per cent of people are happy with that, probably not including those suffering from the disturbance and destruction to crops and gardens that large numbers of them sometimes create. The Forest of Dean harbours probably the largest British boar population, subject to contentious annual culls that some people think too small and others too large. The reappearance of this animal in the British countryside is, at least in part, a consequence of government dithering. The first known escapes, from a farm in Kent, followed storm damage to a perimeter fence in 1987. Year on year, the powers that be debated about whether to try and exterminate them. It seems that no decision one way or the other was ever made, and eradication now would be both difficult and generally unpopular. They are back to stay, and most of us celebrate this return and live with it, as people do across the rest of Europe.

Predator reintroductions are the least favourite propositions. Lynx had 37 per cent support, wolves 36 per cent and brown bears just 25 per cent. Objections to such major changes to the UK fauna are founded on three main concerns: they might spread disease, damage the farming industry and/or eat people. Lynx seems the most likely candidate of these three species to return to Britain in the foreseeable future, and feasibility studies have been underway for some time. It will be essential to consult widely, especially in Scotland, about bringing back any of these charismatic animals, as this is the country best-placed to provide the necessary space. For the record, none of them eat people.

8.2 Overpopulation: An Old Concern

Worries about too many people in the land go back a long way (Foreman, 2011). More than 2,500 years BP, the impacts of human settlers around the Mediterranean roused concerns about desertification due to excessive numbers of goats and to the felling of too many trees. The consequences of that ecological assault persist to this day, with extensive landscapes of rock and scrub generated by those early farmers still surrounding much of the Mediterranean basin. Overexploitation of natural resources was not limited to this geographical region but has continued around the world throughout recorded history. The inspiring Maya civilisation of Central America is a dramatic example of how environmental disaster can strike at the heart of a human society. In the late eighth century AD, the whole edifice of this sophisticated community collapsed, probably due to a combination of unsustainable farming, overpopulation and increasingly limited natural resources. Humans have been slow to learn from events such as this.

In the early days of modern science, clergymen made profound contributions to a growing interest in the world around them. Gilbert White, pottering in his Selborne garden, was one such. Thomas Robert Malthus was another. In his book *An Essay on the Principle of Population* written in 1798, the scene was set for a debate that has continued ever since. The essence of the Malthusian argument is that populations will expand to the point where food supply becomes limiting, and that the potential for such growth will inevitably outrun increases in food availability. His vision was one of future famine and competitive conflict for diminished resources. Malthus was concerned with the future of humanity, but his ideas became central to the most perceptive and far-reaching theory in the history of biology. Charles Darwin 'happened to read for amusement Malthus on Population, and being well prepared to appreciate the struggle for existence which everywhere goes on . . ., it at once struck me that under these circumstances favourable variations would tend to be preserved and unfavourable ones destroyed'. By a remarkable coincidence, Alfred Russell Wallace read Malthus on his sick bed while suffering from malarial fever on a remote Indonesian island and made the same connection. Thus was born the theory of evolution by natural selection.

In the meantime, the Malthusian theory remained firmly in the human domain and made a strong impact on British social policy. It challenged a perception that fertility added to national wealth, and argued that large families were instead often the cause of severe distress. Malthusian ideas were subsequently incorporated into economic theories. Controversy has followed Malthusian economics ever since. He failed to anticipate the impact of contraception on birth rates, which was hardly surprising at a time when effective methods did not exist. And many economists, besotted with the idea that growth can continue indefinitely, stressed that Malthus took no account of technological developments such as the agricultural revolution that can reduce growth constraints. Nevertheless, the limitations on population growth highlighted by Malthus remain a keystone for those now concerned about human numbers in the twenty-first century.

Into the twentieth century, and concerns about unrestrained population growth continued, but mainly in the background and not often visible above the parapet. This changed dramatically with the publication of an influential book, *The Population Bomb* (Ehrlich, 1968). Paul and Anne Ehrlich put together a formidable case for population policies necessary to constrain damaging growth that, within decades, could put devastating pressures on the natural environment and human food supplies. The Ehrlichs gave packed lectures around the world, one of which I was excited to attend in London. Their case was compelling and for a while was taken seriously by leading politicians. US President Richard Nixon recognised the serious challenge posed by population growth in 1969, in a little-remembered statement drowned out by the aftermath of his later resignation in disgrace. Shortly after the Ehrlich book came *The Limits to Growth* (Meadows et al., 1972), a computer-based simulation by the Club of Rome also anticipating dire consequences if economic and population growth was not restrained. The book received harsh criticism from some economists, but many of its forecasts turned out to be close to reality. The Club of Rome persisted, and in a 30-year update asserted that its earlier modelling had proved accurate, perhaps even a shade overoptimistic. The Ehrlichs have continued to campaign on the population issue, but by the end of the 1970s, a huge opportunity to address the problem was lost. In a later work, the authors expressed their frustration at the failure of society to address environmental problems, including overpopulation, despite overwhelming evidence of the urgent need to do so (Ehrlich & Ehrlich, 1996). No doubt many wildlife conservationists share that view.

Optimism about possible policies to regulate population growth peaked by the turn of the 1980s and then faded like a wistful dream. Why so? Several social developments converged in an unlikely and certainly unplanned alliance to turn down the heat on population concerns. The period coincided with a drop in fertility in many developed countries, which was one reason for a change of attitude, leading to utterly misleading claims that the population problem was solved. Not so, of course. Elsewhere around the world there was no such reduction and people numbers continued to rise inexorably. Then came an upsurge in anti-abortion campaigns, led in large part by a Catholic clergy claiming that action for population control was contrary to their religious beliefs and that population increases were not a serious problem in any case. At the same time, members of the rapidly growing feminist movement began alleging that coercion was attempting to restrict the universal rights of women to choose how many children they should bear. Perhaps most significantly, left-wing political activists took against population control, asserting that it was a ploy to blame the world's poorest people for environmental disasters that were caused by overconsumption in richer nations. This attitude has been exacerbated by resistance to increasing levels of immigration, about which any critical concerns have been labelled racist. These features of society have persisted and have successfully dampened down lobbying for population control by prominent conservation organisations. In the early days, the UK Ecology Party (now the Green Party) estimated the numbers of humans that would constitute a sustainable population; Oxfam lobbied for zero population growth; and Greenpeace pleaded for families to 'stop at two'. Unfortunately, in the face of the shifting political mood, such unequivocal stances were abandoned, and for a long time most conservation organisations backed away from population issues altogether. Whether there were sincere changes of mind or just concerns that non-politically correct policies would upset their supporter bases is unclear. However, there have recently been some changes of heart, probably due to an increasing realisation that the fears raised in the 1970s were justified. The Green Party developed a policy on population issues, albeit a weak and convoluted one careful to omit any serious action commitments. Greenpeace also recognised, once again, that population control must be central to long-term conservation goals. Regrettably, Oxfam remained less enlightened. As recently as 2009, the organisation's head of research lambasted a report equating population growth and environmental degradation as a gross oversimplification. Arguably ignoring population problems and dwelling on other, undoubtedly important pressures damaging developing countries is an even worse, irresponsible simplification.

Since the turn of the millennium there has, at last, been a faltering but steady resurrection of a need to address the burgeoning human population. At the forefront of this revival in Britain has been Population Matters, a registered charity dedicated to raising the profile of overpopulation across society. This continues to be an uphill struggle, but progress has been made. Population Matters has enjoyed media coverage and has attracted some prominent patrons including Paul Ehrlich, David Attenborough, Jane Goodall, John Guillebaud, Jonathon Porritt and Chris Packham. All of these people have made public statements about the dangers of overpopulation, and Packham made a televised documentary on the subject in 2020. Population Matters has produced multiple press releases over the years (seven in 2020) and has contributed to discussions on Newstalk and Newsnight, as well as having letters published in daily newspapers. Nevertheless, there is still a long way to go. Most politicians, almost all economists and many environmentalists remain resistant to serious debate about this profoundly important subject. Biodiversity loss and climate change hog the limelight, and it has been dispiriting to note that popular, high-profile television programmes on these important threats to humanity and wildlife still rarely even mention population issues as a major underlying cause. As an elephant in the room, the problem of overwhelming human numbers could scarcely be larger. More like, perhaps, a whale in the fish tank.

Following the increased attention focused on population problems in at least some corners of society, it is pertinent to ask whether people's thoughts about population policies reflect these developments.

8.3 Views on Population in the Twenty-first Century

The population issue crosses all sorts of social boundaries, and it is therefore important to assess how different groups of people feel about it. The most important and certainly the largest such group is the general public, but within that broad remit are various smaller factions. These include naturalists, scientists, artists, religious leaders, economists and politicians. All these people have opinions that will make a difference to future policies on population, and some are supporters of Population Matters.

Public Opinion

Polling has been used several times over the years to gauge public views about population. In a 2018 YouGov poll, 74 per cent of UK adults believed the government should have a national strategy for addressing population, while 64 per cent thought the rate of population growth projected by the ONS is too high. The four main negative effects of population growth were considered to be: impact on public services such as health and education (41 per cent), housing shortages or increases in house prices (15 per cent), impact on the natural environment (9 per cent) and impact on quality of life (9 per cent). There were some substantial differences in response according to political inclination. Among Conservatives, 81 per cent thought population growth was too high. Corresponding figures for Labour and Liberal Democrat sympathisers were 50 per cent and 52 per cent respectively. Most other respondents had no opinion or thought current projections were about right. Hardly anyone in any group considered the figures too low. Environmental issues have become increasingly important to voting intentions, driven primarily by publicity about climate change. Running up to the 2019 UK general election, 25 per cent of those questioned by YouGov put the environment as a major issue on a par with the economy, up from 8 per cent prior to the 2017 poll. Brexit was the dominant concern in both elections for more than 65 per cent of respondents, followed in 2019 by healthcare (38 per cent) and crime (28 per cent). Immigration was of concern to about 20 per cent of people, although overpopulation was not explicitly considered. Polls on population have at best asked general questions about impacts on the environment, without making specific connections to wildlife declines.

Public opinion has also been sounded out beyond UK shores. In an international poll across 17 countries in 2016, only 7 per cent of worries about the planet's future focused on population growth, but Britain was among the most concerned nations on this topic. International terrorism topped the list at 25 per cent, but two other issues (poverty/hunger/thirst, and climate change) gave a combined concern priority of 28 per cent. Evidently, the connections of these problems with overpopulation were not widely made. Things looked different a couple of years later, in a Global Challenges Foundation survey across 10 countries in 2018, where 70 per cent of 10,000 respondents believed that population growth represents a catastrophic global risk and 56 per cent reckoned that the consequences of population growth will be negative for humankind. There were also calls for action. More people thought there should be enforceable international decisions to stop population growth (53 per cent) than did not (40 per cent). An earlier case study in Sweden revealed that 71 per cent of its citizens believed the consequences of population growth to be bad news, whereas the most common answer among politicians was neither positive nor negative (41 per cent). Almost a quarter (23.6 per cent) chose not to answer, and one member of the Swedish parliament thought it was 'an overly complicated issue that is better not addressed here'. Evidently, there can be a substantial dichotomy between the public and their elected representatives on the issue of human population trends. This does not happen only in Sweden.

A great many people from all walks of life have voiced concerns about rampant population growth and its threat to Planet Earth. These are partitioned below into subsets of social groups, and an informative list of those considered here, as well as others, is given on the Population Matters website.

The Naturalists' Perspective

Whenever I talk about wildlife conservation in society meetings, the issue of human numbers frequently comes up at question time. Unfortunately, most people don't make a connection between the two, probably because media attention invariably concentrates on the immediate causes of wildlife declines outlined in earlier chapters. Naturalists, however, are increasingly engaged with population pressures. Peter Scott, one of the earliest well-known British naturalists (Figure 8.2), raised the issue several decades ago in relation to the impact of the WWF, of which he was a founder. His pessimistic conclusion was that it had failed badly in its mission, and that 'if only we had put all that money into condoms, we might have done some good'. In retrospect, this was an unduly harsh judgement on an organisation that continues to do a lot of valuable work, but his point was nevertheless sound and prophetic. Fauna & Flora International is another big player in global wildlife conservation, with a wide and successful remit that improves the

Figure 8.2. Peter Scott, an early advocate of population control to save wildlife.
Source: Fox Photos/Hulton Archive/Getty Images

lot of many species and habitats. Unfortunately, it makes only passing comments on human population pressures, taking the line that it is a problem to be circumvented rather than directly addressed. Another early advocate of population control was the French oceanographer Jacques Cousteau, maker of many astonishing films about the wonders of the undersea world. In 1992, Cousteau's downbeat assessment of human futures foresaw a time when we will be 'living like rats' if our population continues to rise unchecked. David Attenborough's brilliant wildlife documentaries attract huge international audiences and are powerful influencers of public opinion. From one of these, many years ago, a brief final comment about human population problems had to be edited out to ensure acceptance by an American audience. Fortunately, mood music subsequently changed for the better, and Attenborough has since made public statements about population concerns. His overview is that all environmental problems are more solvable with fewer people, and much harder, eventually impossible, to rectify with ever more humans on Earth. Chris Packham has become arguably the best-known naturalist featuring British wildlife on UK television. Packham has not hesitated to make a strong case for action on population control, and discussed the matter with Attenborough on a BBC Horizon programme dedicated to the subject in 2020. And there are others. Kate Humble has made it known that she chose not to have children and that, bearing in mind the world's population size, this was one of the most environmentally friendly things she could do. Jane Goodall's lifetime work studying chimpanzees has been an inspiration to naturalists around the world, combining good science with a passion for conservation. Goodall has expressed the view that current environmental issues wouldn't be a problem if the human population had remained the size it was 500 years ago.

It is impossible to know how many of Britain's thousands of amateur naturalists concur with the importance of human numbers on declines of the wildlife we all enjoy, but hopefully ever more will make that connection and lobby for politicians to address it.

Scientific Concerns

Many philosophers and scientists have expressed concerns about overpopulation. Even in pre-Christian times, Aristotle and Confucius recognised the potential danger of too many people trying to survive with limited resources. Over two millennia on, Bertrand Russell saw the dilemma and opined that 'the one real remedy is birth control'. Physicists were, perhaps surprisingly, early among the scientific fraternity to draw attention to the perils of overpopulation. Albert Einstein considered that it 'has become a serious threat to the well-being of many people and a grave obstacle to any attempt to organise peace on this planet of ours'. The complete list of scientists advocating population control is a formidable one, with a roll call including many of the most celebrated researchers of our time. Ecologist E. O. Wilson, astrophysicist Carl Sagan, and theoretical physicists Stephen Hawking and Max Born are among the best-known names. Emma Woods, head of policy at the UK's prestigious Royal Society, has made it clear that 'we ignore population at our peril'. Chemist David King served as a senior scientific adviser to several UK governments, with an emphasis on climate change policy and, in 2020, on dealing with the coronavirus pandemic.

He stressed that 'the massive growth of the human population through the 20th century has had more impact on biodiversity than any other single factor'. John Guillebaud, Emeritus Professor of Family Planning and Reproductive Health, is a patron of Population Matters and has long campaigned for overpopulation to be taken seriously in discussions about human futures. His take-home message is: 'We must explain to UK couples who plan a family that stopping at a maximum of two children, or at least having one less than first intended, is the greatest contribution anyone can make to leaving a habitable planet for our grandchildren.' In 1994, Guillebaud set up a time-capsule project in Kew Gardens that included environmental concerns of that time, especially overpopulation, which is due to be opened in 2044. The exercise is based on a Kashmiri proverb: 'We have not inherited the earth from our grandparents, we have borrowed it from our grandchildren.' Zoologist Aubrey Manning, another eminent scientist, was also a patron of Population Matters and a passionate conservationist. In a timely anticipation of the importance of access to wild nature emphasised during the coronavirus pandemic, Manning asked many years ago: 'Does everyone now suffer from nature-deficit disorder?'

Norman Borlaug is credited with fathering the Green Revolution, based on the development of high-yielding crop varieties in the 1960s that were responsible for saving a billion people from starvation in the late twentieth century. It was largely due to this highly successful venture that attention was drawn away from the threat of overpopulation. But Borlaug never considered that the Green Revolution would be a long-term solution to feeding the world. He went on to declare that 'the frightening power of human reproduction must also be curbed; otherwise the success of the Green Revolution will be ephemeral only.... Most people still fail to comprehend the magnitude and menace of the "Population Monster".'

Scientific prestige among the general public improved dramatically during the coronavirus pandemic following staggering successes with vaccine development. It must be hoped that this increased respect carries over to scientific proclamations about the risks of overpopulation.

NGOs Involved in Nature Conservation

A bevy of international NGOs exist to promote wildlife conservation. Many are small and specialise in particular groups of species, but some are large and politically influential. Although almost all have taken up the baton on climate change, their collective track record on population issues is dismal. After a promising start with founder Peter Scott's comments discussed above, the WWF backed off further discourse on population problems. This large charity promotes vast amounts of effective conservation all over the world. How disappointing, then, that their Living Planet Report of 2020 barely mentions population pressure at all (World Wide Fund for Nature, 2020). Greenpeace is another multinational actor in the field of wildlife conservation that has studiously ignored population growth through most of its 50-year existence. It is heartening to note, however, that Greenpeace International recently acknowledged that there is a problem that needs to be addressed. Friends of the Earth (FoE) is another big player. It has campaigned successfully on many important issues since its inception in 1969, but for most of its history it had nothing to say about the impacts of population growth. At least it is now on FoE's agenda, but in common with most comparable NGOs, its emphasis remains on the consequences of overconsumption. However, former FoE director Jonathon Porritt and now a Population Matters patron is a vocal advocate of population control.

Based in the UK are several other substantial NGOs with inputs into wildlife conservation. Again, though, population is rarely on the agenda. The National Trust is a major landowner with nature reserves dedicated to protecting a wide range of plants and animals. The Trust has nothing to say about population pressures. Criticism of its failure to highlight population issues clearly touched a nerve in the hierarchy of the Council for the Protection of Rural England (CPRE), an organisation striving to resist ever more encroachment on the UK's green space. A statement by CPRE's Chief Executive acknowledged that population growth is an ongoing problem but went on to list reasons why the CPRE should not become involved with it. The RSPB is a hugely successful conservation charity with many nature reserves and a membership in excess of 1 million. You will search in vain, however, for any mention of human population on the RSPB website. Finally, county Wildlife Trusts have memberships of hundreds of thousands and play leading roles in conservation across the UK. And again, they have nothing to say about population pressures.

The silence of most NGOs on the often devastating impacts of increasing human numbers on their primary objectives in nature conservation is depressing and perplexing. There is an impression that they fear offending, and thence losing the members they rely on for financial support. Yet opinion polls indicate that this fear is unfounded, and it is arguable that leaving the subject in abeyance amounts to a dereliction of duty.

Arts and Media People

It doesn't require a scientific background to become aware of the perils of overpopulation. Many people in the public eye, including those associated with the arts side of life, have voiced alarm about the way things are going. Activist and actor Jane Fonda is one such person, exclaiming that among the many things to worry about in this world, the prospect of 10 billion people living on it was the worst. Others in the acting profession have been similarly forthright. Susan Hampshire, a patron of Population Matters, remonstrated that 'cramming ever more people onto our little planet does it ever more damage – I cannot understand why so many people find this so hard to grasp, and why so many Governments ignore it.' How true, and as Cameron Diaz pronounced, 'We don't need any more kids. We have plenty of people on this planet.' Jeremy Irons concurred: 'There are just too many of us, the population continues to rise and it's unsustainable.' Helen Mirren is one of the UK's favourite actors, and in company with high-profile celebrities Jennifer Aniston and Oprah Winfrey, she considers that 'It is very fine not to want children. There are far too many people in the world. It is my contribution to ecology.' Joanna Lumley, another well-loved actor, remarked, 'I don't think there's any denying the fact that there are too many people in the world . . . and we've never addressed that.'

Musicians have voiced their concerns too. Activist and folk singer Pete Seeger (Figure 8.3), an outstanding supporter of the civil rights movement in the mid-twentieth century, came to recognise that "Some things are so big, like this population problem, that the best way to tackle them is in small ways. I can sing a song about overpopulation and maybe touch one or two people at a time with it." Popular musician Bob Geldof organised Live Aid concerts in 1985 to fund relief from devastating famines in Africa. Evidently, there is a limit to how successful such efforts can be in a situation where population increase constantly outstrips provision, and Geldof later reckoned that 'The tipping point has been reached. There can't be more people on the Earth than we can feed.'

Then there are authors. Isaac Asimov wrote some of the most widely acclaimed science fiction books of all time. He also expressed strong views on overpopulation: 'Democracy cannot survive overpopulation.

Figure 8.3. Pete Seeger, one of many artists recognising a population problem.
Source: Bettmann/Getty Images

Human dignity cannot survive it.... Which is the greater danger – nuclear warfare or the population explosion? The latter absolutely!' Margaret Attwood is a multiple award winner with, among other things, two Booker Prizes to her credit. She is also an environmentalist, and remarked some while back that 'For everybody in the world to have the same lifestyle that we in the West have now, at only six billion people, would take four additional Earths in resources.'

Some outstanding personalities are impossible to categorise. Comedy genius Spike Milligan made contributions to public life way beyond his incomparable humour. Population stress was one of many issues that took his fancy, on one occasion announcing that 'Overpopulation is a serious issue. The human race will soon have to get used to 12 in a room.' Milligan was a Catholic, once writing to a Father Patrick Fury that 'One day, dear Patrick, when the Vatican is filled with the beds of overflowing families our dear Pope might shout out enough is enough...and at communion not only give out the Host, but also the pill.' This correspondence apparently terminated their friendship. A damning reply might have mentioned that Milligan fathered five children. Monty Python stalwart and world traveller Michael Palin announced his view that 'The greatest politically charged challenge facing our planet? Unchecked population growth.' Gloria Steinem is a renowned journalist and political activist and recognised as a leader of the American feminist movement in the late mid-twentieth century. In a pointed appraisal of how women might consider reproduction, she suggested that 'Everybody with a womb doesn't have to have a child any more than everybody with vocal chords has to be an opera singer.' Adrian Hayes is a polar explorer and adventurer, and also a patron of Population Matters. His take is like so many others listed in this chapter: '. . .it is pointless campaigning against climate change or to "save the Arctic" without addressing the root cause behind it and virtually every other environmental issue we face: our unsustainable numbers on this planet. That is the real "inconvenient truth".'

Many more people than those listed here have expressed deep concerns about overpopulation. This section is just a sample of much concordant thought on the subject.

Religious Views

Antipathy of the Catholic Church to artificial methods of birth control is well known but is not typical of other religious groups around the world. Acceptance of contraception by any faction of the Christian church is relatively new, and all variants of this religion disfavoured the practice until the start of the twentieth century. However, the Church of England has long approved of contraception although its relevance to population issues, rather than to individual freedoms, is rarely mentioned. George Carey, a former Archbishop of Canterbury, considered that 'The overpopulation of this small island nation, already stricken with a mountain of debt that could blight generations, is the gravest crisis we face.' Martin Luther King, a Baptist minister and heroic activist, equally recognised the dangers of human overload: 'The modern plague of overpopulation is soluble by means we have discovered and with resources we possess. What is lacking is not sufficient knowledge of the solution but universal consciousness of the gravity of the problem and education of the billions who are its victims.' By and large Protestant opinion is accepting both of birth control and that population issues need to be addressed. On the fringes, however, some Evangelical wings of the Christian church maintain that action on these activities, as well as on other green issues such as climate change, are contrary to biblical teachings.

Hinduism is the dominant religion of India and does not ban birth control. The subject is not a major ethical issue there, and some Hindu scriptures even provide contraceptive advice. Because India has such a large population, much discussion of birth control has focused on the environmental impacts of overpopulation rather than on more personal ethics. Unfortunately, the subcontinent has an anguished history with respect to birth-control policies. India was the first country in the world to implement, in the 1950s, a state-sponsored family planning programme. In a dramatic development in the early 1970s, Prime Minister Indira Gandhi invoked forced sterilisation for men with two children or more. The plan failed and unsurprisingly became very unpopular. It is still remembered and widely blamed for creating a public aversion to family planning, which hampered subsequent government programmes for decades. A recently proposed Population Control Bill has also generated controversy because it is considered Islamophobic by Muslims, who resent allegations that their community is driving population growth in India. This part of the world is especially needy of a consensual approach to this vitally important topic.

Islam accepts the use of contraception. According to an Islamic viewpoint, the number of children should be 'as many as they do not cause any harm to parents, while at the same time parents would provide all their juridical, legal, spiritual, cultural and educative rights'. So family planning as the means of regulating the family size for health or economic reasons is permissible in Islam, and some Muslim countries, such as Iran, have advocated its use.

Buddhism is also accepting of the need for population control. The 14th Dalai Lama, spiritual leader of the Tibetan people, has expressed serious worries about it: 'One of the great challenges today is the population explosion. Unless we are able to tackle this issue effectively we will be confronted with the problem of the natural resources being inadequate for all the human beings on this Earth.' Enlightened views are therefore the norm among the majority of world religions.

Roman Catholicism is in the minority with respect to religious attitudes to birth control but remains an obstacle for attempts to curb population growth in countries where that is the dominant religion. There are many such countries, especially in Central and South America, where the doctrine has caused misery by forcing a dichotomy between religious teachings and poverty consequent upon overlarge families. While there has been some limited acceptance that overpopulation is a problem that frequently exacerbates such poverty, there has been no movement on Papal encyclicals of the 1960s forbidding all but the 'rhythm method' contraception. Pope Paul VI conceded that 'There is no denying that the accelerated rate of population growth brings many added difficulties to the problems of development where the size of the population grows more rapidly than the quantity of available resources to such a degree that things seem to have reached an impasse. . . .it is for parents to take a thorough look at the matter and decide upon the number of their children.' Half a century later, Pope Francis is also cognizant of environmental deterioration and the need to act on it. In 2015, he issued an encyclical that warned of impending global disaster because of climate change. A Vatican workshop in 2017 on 'How to Save the Natural World' brought international experts, including Paul Ehrlich, together to discuss policies including population control. Unsurprisingly, this development made many conservative Catholics unhappy and the encyclical prohibiting artificial birth control still stands. In practice, however, Catholics have now overwhelmingly rejected the encyclical's teaching. A 2014 poll found that large majorities of Catholics in traditional strongholds of the faith favoured the use of contraceptives: 93 per cent in Brazil, 84 per cent in Italy and 68 per cent in the Philippines. In the USA, a 2016 study by the Pew Research Center found that only 13 per cent of weekly Mass-going Catholics thought contraception was morally wrong. Perhaps events on the ground will eventually drag the Vatican into a more credible position on birth control, but it is a perilously slow process.

Economists and Population Issues

Economists have been a major bugbear for those trying to promote population control. The rather few media debates on population issues regularly include economists as antagonists to action on the subject. It is therefore unfortunate for population lobbyists that the second most senior post in the UK is always an economist, in the form of Chancellor of the Exchequer. It is pertinent to ask why economics is so highly rated in the corridors of power.

Scottish philosopher Adam Smith is credited with founding modern economics in the eighteenth century. Several academics subsequently explored his ideas, focusing mainly on so-called microeconomics: how markets work, and how people make choices based on self-interest about buying and selling, all in ways that relate to individual livelihoods. A more general theory by John Maynard Keynes in the early twentieth century expanded the subject into macroeconomics, an attempt to describe and ultimately influence how economies function at national and international levels. This requires consideration of many factors including levels of unemployment, interest rates and inflation, and proposes that government intervention can influence all these familiar news items for the public good. A basic goal is to increase wealth by continuous economic growth, manifest as increases in the gross domestic product (GDP). The obsession with economic growth and ever-increasing per-capita GDP is quite new and was not clearly articulated until the late nineteenth century. There was already by then a paradigm that economic growth is a 'good thing', but the first half of the twentieth century saw its definition sharpen. This foreordained an emergent conviction that promoting growth is a matter of national priority. By mid-century, GDP growth was firmly established everywhere and for a long time was widely assumed to be the route to prosperity. Since then it has become apparent that continuous GDP growth leads not simply to wealth and well-being but potentially to environmental collapse and ecological disaster. It is therefore around the focus on GDP growth that conflicts with environmental protection, including population control, have surfaced most strongly.

Unfortunately, reservations about growing GDP as the Holy Grail of modern society have passed most economists by, including chancellors. Even demonstrable dangers such as climate change have, until recently, fallen on deaf ears and are still denied by at least one former Chancellor of the Exchequer. A common argument put forward by economists is that constraints on birth rates invert the 'demographic pyramid', which in most species, including humans, has a wide base with lots of youngsters, declining upwards with a few oldies at the top. In upside-down mode, a large elderly population must be supported by relatively few young people. Much the same case is put forward to maintain immigration of extra workers. The case fails on simple logic. It requires indefinite increases in population size, an obviously impossible scenario. But zero-growth, 'green' economies have been envisaged. Increased use of technology, maybe with targeted tax increases, can maintain a temporarily inverted pyramid, which will eventually revert when a new equilibrium of birth and death rate is reached. Increasing alarm about climate change and other environmental threats has generated a 'degrowth' movement (Figure 8.4), calling on rich countries to embrace zero or even negative GDP growth. Giorgos Kallis, an ecological economist at the Autonomous University of Barcelona, emphasises that 'There is no

Figure 8.4. The case for halting eternal economic growth. *Source*: Meteor Independent Media, https://themeteor.org/category/editorial/ (A black and white version of this figure will appear in some formats. For the colour version, please refer to the plate section.)

way to both have your cake and eat it, here. If humanity is not to destroy the planet's life support systems, the global economy should slow down.' Yet most economists remain ensconced in a self-propagating bubble that ignores events in the real world surrounding them. Habitat loss and biodiversity crises are not their concern. On the rare occasions when their position on impacts of population increase on world events are challenged, their responses typically include reference to population growth slowing down, ignoring the wide variation on this prediction discussed in Chapter 7. Even if true, no account is taken of the environmental damage already underway with a population that has even now exceeded safe limits. Economists also make occasional, ludicrously huge predictions about how large the human population could safely become. Sadly, the increasingly discredited goal of continuous economic growth is still perpetuated uncritically in the media. We are regularly informed about percentage growth rates both local and national, and comparisons are frequently made with other countries that are 'doing better'. Negative growth can become a 'recession', an emotive term calculated to indicate that we must up our game.

Not all economists are environmental philistines. Nicholas Stern produced an inspired report on the economic perils threatened by climate change, and former Bank of England governor Mark Carney made similar comments. According to Kenneth Boulding, President Kennedy's Environmental Adviser in the 1960s, 'Anyone who believes in indefinite growth of anything physical on a physically finite planet is either a madman or an economist.' Robert McNamara, former President of the World Bank reckoned that 'Short of nuclear war itself, population growth is the gravest issue we face. If we do not act, the problem will be solved by famine, riots, insurrection and war.' And economist Partha Dasgupta, another Population Matters patron, recognised that 'Population growth, poverty and degradation of local resources often fuel one another.' In a comprehensive report on the importance of biodiversity in economics, Dasgupta describes the benefits of readily available family planning among a wide range of other features that economists would do well to take seriously (Dasgupta, 2021). We need more economists in high places with this frame of mind. A good case can be made for environment ministers to have a status in government at least as high as that of economists. This would go some way to counter the overarching but unwarranted importance of economics in the governance of our country. It is not as if economists have an impressive track record in forecasting economic trends. Economics Nobel Laureate Friedrich Hayek made a frank admission nearly half a century ago, stating that not only were economists unsure about their predictions but that presenting their findings with scientific language was misleading and 'may have deplorable effects'. Forecasting accuracy has improved since then, but Prakash Loungani's analysis at the International Monetary Fund found that 'The record of failure to predict recessions is virtually unblemished. Economists failed to predict 148 of the past 150 recessions.' As a subject, reappraisal of its track record and demotion of economics as a major driving force in the political arena might well be the best of news for the future of Planet Earth.

Politics and Population

After a heyday in the 1970s, politicians of all parties in the UK have generally shied away from discussion about population problems. Americans have been more vocal in commenting about population issues. Not only Richard Nixon, but previous US president Lyndon Baines Johnson also recognised the difficulties posed by increasing human numbers: 'The hungry world cannot be fed until and unless growth of its resources and the growth of its population come into balance. Each man and woman and each nation must make decisions of conscience and policy in the face of this great problem.' Later on, American vice president Al Gore became the most prominent US campaigner on environmental issues, most notably for his attempts to make the world take climate change seriously. He has taken a similarly enlightened stand on population pressures: 'Population growth is straining the Earth's resources to the breaking point, and educating girls is the single most important factor in stabilising that. That, plus helping women gain political and economic power and safeguarding their reproductive rights.' Environmental issues in the USA remain a hard sell, however, in a country where scepticism about climate change remains rife and where almost 40 per cent of the population adopt creationist beliefs rather than accept evolution.

In recent times, it is both the far right and far left that have commented most frequently on proposals to curtail population growth, always in a negative vein that sometimes borders on the phobic. A combination of emotive language and red herrings typify their arguments, and neither group seems to accept that population growth has already generated huge environmental damage. Alexandra DeSanctis tells us that 'The Left has long been infected by a dangerous desire to tell you how many children you're allowed to have. ...If progressives plan to continue indulging this latest disturbing strain of population-control rhetoric, they ought to be forced to confront the question: Who among us is less worthy of being alive?' The main concern in this case seems to centre around an infringement of human rights, and particularly the choice of how many children people should have. Most of those on the far left would no doubt be astonished to learn that they were anything other than damning of population control proposals. The Socialist Workers Party entreats us 'not to buy Chris Packham's myth of overpopulation' and asserts that Population Matters is a right-wing pressure group, which as a registered charity it certainly is not. The core argument is that population control is directed primarily at poor countries (it isn't), and diverts attention from unequal levels of consumption, with rich countries taking a disproportionate amount of the world's resources (which is true). As outlined in this book, overpopulation is at least as much a rich as a poor country problem and has no racist overtones. Control of human numbers is ultimately in everyone's interest, rich and poor. It should, however, be contemporaneous with a long-needed redistribution of wealth across the globe.

What about mainstream political parties in Britain? Tories have raised the issue of population control from time to time, including Prime

Minister Boris Johnson and his father. In 2015, Stanley Johnson published his views on why Britain needs a population policy, and in a 2007 newspaper article, Boris Johnson opined that climate change is not the biggest challenge faced today but that the reproduction of our species is a far bigger one. Needless to say these pronouncements have not translated into policy statements, let alone decisions, and cynics might relate the Johnson suggestions to the numbers of children that both men have fathered. Look in vain for anything to do with population issues on the Labour Party website, which has plenty of otherwise heartening plans for environmental protection and wildlife conservation. The same applies to Liberal Democrat green policies, which ignore population altogether. We might reasonably expect the Green Party to take a stand on this subject, but as mentioned earlier, its stance on population is vague and full of waffle (although at least it has one), and seems designed to minimise offence rather than to propose action. However one-time Green Party member Sara Parkin has championed the need for population control, and as a patron of Population Matters reminds us and her former colleagues that 'it is no longer legitimate to leave policies for lowering birth rates off the policy agenda'.

This is disappointing news, and surely reflects a disconnect between politicians and the views of those electing them. However, at last there are signs of a political awakening. In *Overcrowded Islands?*, the Institute for the Study of Civil Society (Civitas) provides a comprehensive review of population issues in the UK (Hodgson, 2020). This think tank proposes the creation of a Demographic Authority of comparable stature to the ONS in order to consider future aspects of population change. Civitas is not, however, generally progressive in its attitude to green issues. How much more desirable would be a multiparty effort to consider future prospects for the UK's human population size.

8.4 Other Consequences of High Population Density

Declining wildlife is only one of the problems consequent on a high human population density, and for most people it is probably among the least obvious. Among pressing concerns, health and well-being stand out as significant casualties of overpopulation. In 1972, when population issues were freely discussed and had not yet become a taboo subject, 50 medical doctors published the same letter in *The Lancet* and the *British Medical Journal* (itself a unique achievement) urging politicians to develop a population policy specifically for the UK. They expressed concern about how overcrowding in cities with its attendant pollution was a 'direct threat to the mental and physical wellbeing of our patients'. Disappointingly, the proposed promotion of family planning and improved career prospects for women went largely unnoticed by the government. Undaunted, the medical concern about too many people did not go away, and a 'doctors and overpopulation' group has persevered with drawing attention to the topic, recently reiterating the need to act on it (Gregus & Guillebaud, 2020). What, then, have been the most important downsides of population increase to quality of life in the UK?

Transport Difficulties

Traffic jams are an obvious manifestation of overcrowding in modern Britain (Figure 8.5). They have been with us almost as long as the motor car. I still remember, as a child in the 1950s, sitting for what seemed like a lifetime on the Exeter bypass en route to a summer holiday. Few would dispute that congestion on the country's roads has become increasingly awful as time goes by. There are some dreadful extreme instances. On 4 August 2017, traffic tailed back 35 miles at its peak on the northbound

Figure 8.5. An all too familiar traffic jam – just too many people. *Source*: Tim Graham/Getty Images

M5, with the gridlock lasting for 15 hours. There were over 1.35 million traffic jams reported in the year up to October 2017, costing drivers an estimated £9 billion in wasted time and fuel. This is not because individuals are driving more miles each year; rather, as recounted in Chapter 4, the cause is an increasing number of drivers.

The costs are not just economic. Rush-hour commuters in the USA lose an average of 42 hours per year to traffic delays caused by clogged highways. On the nation's 10 most gridlocked roads, the number doubles to 84 hours. That is 3.5 days per year of sitting in congestion. A British study found that blood pressure increased during traffic jams, especially when people listened to aggressive music or to no music at all. Exposure to daily road snarl-ups can lead to high, chronic stress levels. One stress trigger while driving in dense traffic is impatience. This tends to turn into resentment, aggressive driving and anger, which can all lead to road rage. Stress is a killer disease that makes people vulnerable to other disorders, including depression. Regular exposure to traffic jams evidently has significant adverse consequences for human health.

Using public transport is also stressful as overcrowding increases. This anxiety can spill over into commuters' work and home lives. Elevated tensions during journeys contribute to serious health problems including cardiovascular disease and suppressed immune functioning. Regular commuters frequently suffer heart palpitations, higher levels of agitation stress and a sense of lack of control.

Stress associated with flying is, according to a recent traveller survey, primarily due to concerns about missing a flight due to traffic jams en route to the airport. A study by the City Pair Program found that a third of people who fly now believe the airport experience is more stressful than the working week, with nearly a quarter saying it is worse than moving house. Overcrowding is set to intensify the problem. Department for Transport figures revealed that Heathrow, Gatwick, and Stansted airports will be at full capacity within the next 20 years, meaning longer queues, even more crowds and fewer seats, all of which is more bad news for our communal well-being.

Travel Infrastructure

Another consequence of a growing population is the need for more roads, railways and airport facilities. Such developments inevitable impinge on people living in their way, or close to them. Highways England unashamedly promotes extensions of the road network as a route to increased economic growth, despite long-standing reservations about the impacts of ever more traffic on people and the environment. Objections to Highways England's plans rarely succeed in modifying let alone stopping whatever they have in mind. In the 1960s, the new M60 motorway was driven through our previously peaceful Lancashire village and people lost their homes. Now, not so far way, a major new trunk road has been approved to run through Rimrose Valley Country Park in Liverpool. This goes against the wishes of local authorities, while campaigners describe the plan as a declaration of intended vandalism and assert that the government should be ashamed of taking the last remaining decent, open green space away from the people of Sefton. On the basis of previous experience, Highways England seems all too likely to get its way. There are many other examples of road developments that, if allowed to go ahead, will damage quality of life and living space in our green and pleasant land.

Railway extensions in the twenty-first century are something of a novelty after decades of decline, but the devastation under way during construction of the HS2 high-speed rail link between London and Birmingham changes all that. Almost 900 homes, 1,000 businesses and around 60 'irreplaceable' ancient woodlands will be destroyed by the construction of the HS2 railway, according to the company behind the project. The cost of reducing journey time between the two cities by about 30 minutes currently runs at around £50 billion. The economic case for HS2 development looks increasingly weak, and the whole project bears the signature of a political vanity project, in which case, halting the appalling damage to people and wildlife in its wake was never likely to happen.

Then there is the vexed question of airport expansion. Despite climate change demanding fewer rather than more flights around the world, pressure persists to move in the opposite direction. The proposal for a third runway at Heathrow epitomises this dilemma. If it goes ahead, building the new runway will involve diverting rivers, moving roads and rerouting the M25 through a tunnel. Opponents claim that despite assurances to the contrary, the development would mean unacceptable levels of noise and pollution and add to carbon emissions from the increased number of flights. On top of that, 761 homes are expected to go, including the entire village of Longford. Not long ago a British politician promised to lie down in front of the bulldozers if the plan was approved. We will have to see whether the elevation of this individual to Prime Minister affects the outcome of his attention-seeking pledge.

Public Services

Important aspects of life in the UK are supported by long-standing public services. Among these, the National Health Service (NHS) and universal education are prominent in terms of scale and cost. In 2020, the NHS employed about 1.6 million people across the UK, while in 2019, the state-funded education sector employed just under 1 million teachers and support staff. These public servants have had to cope with ever more patients and students as the human population increases. The NHS workforce is growing but not sufficiently to cope with the increasing demand. Between 2005 and 2020, NHS employment expanded substantially with a 44 per cent overall increase in hospital medical staff. Even so, the total number of permanent, qualified general practitioners (GPs) in 2019 was down by 6.5 per cent compared with 2015. In 2020, there were unfilled vacancies for about 8 per cent of the hospital workforce, and these were most acute in the nursing profession. Over the last few decades, demands on the health service have risen dramatically. This can be attributed to a growing and ageing population, as well as to advances in medicine and technology that have enabled a wider range of healthcare services to be provided. The UK has relatively few staff in key health-worker groups compared with other developed countries. Per head of population, the UK has fewer than half as many nurses as Norway (8 nurses compared to 18) and is among the lowest in the developed world in its numbers of doctors. The level of demand for healthcare is expected to continue increasing. The population in England is certain to grow, adding further stress to the NHS for the foreseeable future. All of this has consequences for patients. Even before the 2020 coronavirus pandemic, parts of the NHS in England experienced the longest waiting times since targets were set. This included the highest proportion of people waiting more than 4 hours in accident and emergency departments since 2004, and the highest proportion of people waiting over 18 weeks for non-urgent (but essential) hospital treatment since 2008. Survey evidence also suggests that more people are experiencing lengthening delays in getting GP appointments. To achieve even the current level of service, foreign labour has been essential. The proportion of migrants working in the NHS varies across staff groups and different regions, but in 2019, 13.3 per cent of NHS staff in hospitals and community services in England were of a non-British nationality. Among doctors, the proportion was 28.4 per cent. Many doctors have trained abroad. In 2019, 20.1 per cent of GPs in England qualified outside the UK, and in 2015, about 15 per cent of UK nurses were not British nationals. An increasing population combined with fewer migrant staff paints a gloomy picture for future healthcare in the UK.

Numbers of students in UK schools vary mostly according to changes in birth rates, resulting in increases or decreases years down the line from significant variations in fertility. Immigration of pupils born outside the UK has a very small effect on the school age population, but the birth rate is influenced by any increase in the number of children born to non UK born women, who overall tend to have higher fertility rates. The nursery and primary school population has been rising since 2009 and reached 4.64 million in 2018, the final year with available data. However, the rate of increase is slowing, and the early school population is projected to stabilise in 2019 at 4.66 million before starting to fall. The secondary school population rose to 2.85 million in 2018 and is projected to continue increasing until around 2025, reaching an estimated 3.28 million. These projections could be substantially off course if there are unpredicted changes in fertility or net immigration rates. There is an ongoing shortage of qualified teachers to deal with increases in student numbers. International migration contributes substantially to school workforces, with around 12 per cent of school staff in England born outside the UK in 2015–2017.

Immigrants have, over many years, made hugely valuable contributions to public services in the UK. The country simply could not have functioned properly without them. Inevitably, they are also part, albeit a small one, of the country's overall population increase. Politicians occasionally query why home-grown labour cannot fill the public service vacancies, thus reducing the demand for immigrants. These people might like to look at pay scales. Nurse remuneration in the UK is low relative to many other countries, although comparisons are difficult because working conditions vary a lot among nation states. However, the top 10 payouts come from Luxembourg, Iceland, the USA, Australia, Norway, Israel, the Netherlands, Switzerland, Ireland and New Zealand. The UK is not among them. It is surely telling that British nurses have in some instances resorted to food banks to support themselves. Starting salaries for teachers in the UK are 19th in the list of European countries, and the lowest of any west European state. NHS and teaching staff are among the most dedicated, hard-working and skilled professionals in the land. Most work for love of the job, but persuading more UK nationals to join them would undoubtedly be helped by providing pay scales commensurate with the terrific services these people provide.

Air Pollution and Population

If you are unlucky enough to live in a British city, you are taking serious risks with your health. Air pollution in and around towns and cities is one of the country's biggest killers. Astonishingly, to me at least, between 28,000 and 36,000 early deaths, equivalent to more than 8 per cent of all deaths each year, are caused by air pollution in the UK. The health impacts of exposure to air pollution are long term, and manifest as increased respiratory and cardiovascular diseases. A toxic mixture of sulfur and nitrogen oxides and particulates are the main causes. A landmark coroner's report found that illegal levels of air pollution in London contributed to the death of 9-year-old Ella Kissi-Debrah, who lived near one of the capital's busiest roads (Laville, 2020). About 80 per cent of urban NO_2 air pollution comes from road transport. The increase in road traffic over recent decades has not only impacted the air quality of our towns and cities but has also had wider effects, including excessive noise pollution. Two hundred and forty-eight UK hospitals and 2,220 General Practices are in areas where air pollution is above the safety limit of the World Health Organisation (WHO) for fine particulate matter. The EU Court of Justice ruled that the UK has systematically and persistently broken legal limits on toxic air pollution for a decade. Levels of NO_2, mostly from diesel vehicles, remain illegally high in 75 per cent of urban areas. The case began before the UK left the EU, but the legal limits remain in UK law.

As outlined in Chapter 4, the high levels of traffic across the UK directly reflect the increasing numbers of people driving, and therefore population growth. Measures are underway to reduce air pollution in Britain, and may eventually allow everyone to breathe freely again, but there is a long way to go. As with water pollution, the effort and resources required to make us safe from noxious chemicals are bound to go on rising as long as the numbers of polluters – us – also go ever upwards.

8.5 Overview

Opinion polls confirm that rich wildlife is a widely appreciated feature of the British countryside, with public support for radical measures including reintroduction of lost species. The issue of human population size has a long and complex history with respect to public opinion, surfacing quite dramatically in the 1970s but subsequently dropping off the radar until another recent but gradual revival in the twenty-first century. The general public, naturalists, scientists and many on the arts side of society have increasingly coalesced around the need for a population policy in Britain. Wildlife NGOs have too often remained disappointingly reticent to commit to this idea, as have some (but not by any means all) religious groups. Unfortunately, economists and politicians, the key decision makers, have remained stubbornly recalcitrant and lag behind mainstream public opinion on the need for control of population growth in the UK. Most people probably do not make a connection between human numbers and wildlife declines but are increasingly aware of other downsides consequent on high population density. These include transport difficulties, especially traffic jams, ever more pressures for development, overcrowding on public services and the damaging reverberations of air pollution. The question hangs in the air as to whether British politicians will ever lose their blinkered attitude towards a problem that, as summarised so well by Shragg (2015), underpins many difficulties that range across the whole of society.

9 · *International Aspects of Population Growth*

9.1 Global Impacts of Overpopulation

The focus of this book is on population pressures in the UK, and specifically on that country's wildlife. This, however, should be viewed in a broader perspective. Population growth has been an international concern for decades past. Its effects on wildlife have largely been considered in developing countries, especially in tropical regions of Africa, South America and South-East Asia. This has often entailed an understated hypocrisy. As outlined in earlier chapters, the developed countries of Western Europe frequently have higher human population densities than occur in other continents. Encroachments into the majestic, marvellous habitats of open savannahs and tropical rainforests merely mimic the devastation wreaked on temperate woodlands centuries ago in Europe. This is not to belie the importance of wild lands now under massive threat in the developing world, but attempts to arrest the ongoing destruction there must be pursued in ways sympathetic to the human inhabitants. The examples listed below for three continents are exactly that, just a sample of the many precious places suffering progressive devastation around the globe.

Global concern about overpopulation has been more broadly based than its consequences for wildlife. Its impacts on the standard of living and quality of life of people in many countries have come under scrutiny. International organisations have pontificated on the issue, and there have been multinational conferences on the subject. Unfortunately, agreements about what needs to happen, and how to restrain population growth, are thin on the ground.

9.2 Population Pressures on Tropical Environments

Africa

East Africa is a landscape of grassland plains running on seemingly forever, where some of the world's most charismatic wild animals flourish. Vast herds of wildebeests (*Connochaetes gnou*) pound across the savannah, in the company of impala (*Aepyceros melampus*), buffalo, giraffes (*Giraffa* spp.), elephant (*Loxodonta africana*) families, lumbering rhinoceroses (*Diceros bicornis*) and other instantly recognisable large mammals. We never tire of watching carnivores hunting these huge herbivores; crocodiles (*Crocodylus niloticus*), lion prides, leopards (*Panthera pardus*), hyenas (*Crocuta crocuta*), wild dogs (*Lycaon pictus*) and cheetahs (*Acinonyx jubatus*) all provide remarkable spectacles, usually with an unpleasant ending for the coveted prey. For youngsters like me in the 1950s, East Africa was magically portrayed by the pioneering, eccentric film-makers Armand and Michaela Denis. Armand would ride on the back of rhinos, and Michaela was insistent about applying make-up, assuring viewers that she wouldn't think of getting into the water with crocodiles until she had used her eyebrow pencil. Nowhere else on Earth is large-animal biodiversity so rampant and so easily seen, attracting huge numbers of tourists every year. It remains an inspiring place to visit, but this bountiful arena is not what it was. Poaching, mostly of elephants and rhinos, has seriously depleted these most emblematic of animals, but the main, incessant problem has been people pressure.

Serious damage to the savannah has accrued consequent on climate change, farming practices involving overgrazing, aggressive irrigation lowering the water table, deforestation and erosion. Each year, more than 46,000 km^2 of African grassland are desertified. By 2012, about 75 per cent of Africa's savannahs, and more than two-thirds of the lion population once estimated to live there, had disappeared over the previous 50 years (Riggio et al., 2013). According to one of the authors, 'The word savannah conjures up visions of vast open plains teeming with wildlife. But the reality is that massive land-use change and deforestation, driven by rapid human population growth, has fragmented or degraded much of the original savannah. Only 25 per cent remains of an ecosystem that once was a third larger than the continental United States.' Within this precious landscape, the Serengeti Mara is one of the largest and most protected ecosystems on Earth, spanning 40,000 km^2. Every year, a million wildebeest, half a million gazelle and 200,000 zebra migrate from the Serengeti park in Tanzania to the Maasai Mara reserve in Kenya, searching for water and grazing land. In 2018, a report concluded that park boundary areas had experienced a 400 per cent increase in human population over the past decade and that the wildlife population on the Kenyan side had decreased by more than 75 per cent. Between 1999 and 2012, the human population grew by an average of 2.4 per cent per year, and within 13 years the population had increased by a third. Alongside the people, cattle herds around the Serengeti rose by an average of

0.9 per cent, and sheep and goats by 3.8 per cent. The border between the Serengeti National Park and the agricultural landscape outside it is now very evident (Figure 9.1). Cattle trails fan out as they reach the protected area, as local people illegally graze livestock in there every day. This leads to another conflict, the predation of livestock by lions. In one instance, lions killed a cow during the night and the next day the carcass was poisoned by a herder in revenge. When the lions returned the following night to finish their meal, they were poisoned. An adult female and a young male died, as well as six critically endangered white-backed vultures (*Gyps africanus*). Encroachment into the reserve by cattle is reducing the grass available for wildlife such that the

Figure 9.1. A distinct border between the Serengeti reserve (left) and agricultural land (right). © Daniel Rosengren (A black and white version of this figure will appear in some formats. For the colour version, please refer to the plate section.)

wildebeest migration now routinely spends a month less in Kenya than it used to a decade ago, and most other large-animal populations are falling.

Problems from human population increases in Africa are not confined to the eastern grasslands. The Sahel region is a 3 million km^2 semi-arid belt of barren, sandy and rock-strewn land running across the breadth of Africa, from the Atlantic Ocean to the Red Sea. It forms a boundary between tropical forest regions to the south and the Sahara desert in the north. Despite its arid nature, within historical times the Sahel supported rich wildlife assemblages not dissimilar to those in East Africa. including large populations of the scimitar-horned oryx (*Oryx dammah*), dama gazelle (*Gazella dama*), dorcas gazelle (*Gazella dorcas*), red-fronted gazelle (*Gazella rufifrons*) and Bubal hartebeest (*Alcelaphus busephalus*), along with large predators such as wild dogs, cheetahs and lions. Once again, the larger species have been greatly reduced in number by overhunting and competition with livestock.

The Sahel also constitutes a stopping-off point for many migratory birds, more than 2 billion of which travel to Europe every year to breed. These include many popular but declining visitors to the UK. It is clear that Sahel rainfall has a major impact on migrant bird numbers (Adams et al., 2014). Precipitation in the region is highly variable between years and has been linked to survival for at least 13 migrant species, including birds that overwinter in the Sahel but also others such as nightingales and swallows that just pass through. There have also been profound changes in land use in recent decades. Agriculture has extended onto previously uncultivated land and has become more intensive with shorter fallow periods and increased use of pesticides and irrigation. Areas of forest have been intensively cut for fuel wood and for livestock browse. Forest and grazing reserves have been used more intensively, and a disastrous consequence has been serious desertification of the region. Around 2,500 km^2 are lost to this process each year in Niger alone, an area about the same size as Luxembourg.

Given the nature of the Sahel, living there has never been easy. Unfortunately, the region is experiencing a phase of human population growth unprecedented anywhere else in the world. Over the next 20 years, the total population of five Sahel countries (Burkina Faso, Chad, Mali, Mauritania and Niger) is set to double and reach up to 160 million inhabitants by 2040. It is hard indeed to see how this increase will be sustainable without mass emigration and even more damage to an inherently fragile environment.

South America

Stretching from the equator to the Southern Ocean, with grassy plains, high mountains and tropical rainforests, South America sustains a greater range of plants and animals than any other continent. Its vast tropical forests have attracted innumerable European explorers and have witnessed the demise of many of them, including one Colonel Percy Fawcett who disappeared in 1925 searching for a mythical lost city. The Amazon basin is undoubtedly South America's best-known ecosystem, but others of repute include the Pantanal wetlands, the high Andes pastures and the pampas grasslands. All of these hugely important habitats are under threat. The Amazon rainforest is of international importance both for its biodiversity and as a carbon sink mitigating climate change. It covers a remarkable 5.5 million km^2, or at least it once did. Since 1978, about 1 million km^2 have been destroyed, mostly by deforestation, and the rate of attrition was still accelerating in 2020. This damages not only wildlife but also the indigenous peoples of the region who have suffered enormously from the destruction. A growing human population beyond the rainforest has been a partial cause of forest removal, by a combination of logging, agriculture, cattle ranching, mining, oil extraction and dam building. To the south, the Pantanal, an area of around 200,000 km^2, is the world's largest tropical wetland and also its largest area of flooded grasslands. Pantanal wildlife, easier to see than that in the dense rainforests, is truly wonderful and attracts increasing numbers of visitors. Jaguars (*Panthera onca*) are a star attraction, but there are many other animals to behold. At present, much of the Pantanal ecosystem remains intact, but threats are mounting. Less than 3 per cent of the habitat lies within protected areas, and the region has already lost approximately 50 per cent of its original cover. More than 40 per cent of the forest and savannah habitats have been altered for cattle ranching through the introduction of exotic grasses. Ranching remains the dominant land use, but arable land is expanding over areas historically used as pasture. Economic development and consequent population growth pose a new threat to the Pantanal due to their negative environmental associations. Roads and electricity lines have been constructed, while large agri-industrial projects have emerged on the Pantanal's periphery, all fostering more population increases.

Most of the Amazon and Pantanal regions lie within Brazil, a country in which population growth has slowed substantially since the 1970s. This is good news, and together with measures reducing land clearance for growth of soya beans and cattle, this trend could pave the way for long-term conservation across the region. Unfortunately, in recent years, illegal encroachment, including deliberate burning of rainforest, has revived major concerns about the Amazon's future, but at least in this instance human population pressure may begin to recede rather than further exacerbate the problem.

South-East Asia

The bountiful jungles of Indonesia are a treasure trove of rich biodiversity, sufficient to inspire Alfred Russel Wallace's influential theory on island biogeography and his co-discovery with Charles Darwin of evolution by natural selection. Wallace was way ahead of his time in recognising the universal qualities of humankind, writing to a friend that 'The more I see of uncivilised people, the better I think of human nature, and the essential differences between civilised and savage men seem to disappear.' In Wallace's lifetime, the island archipelago supported 1.7 million km^2 of rainforest in 1900, but a century later only 1 million remained. Indonesia has suffered from the greatest rate of deforestation anywhere in the world, and continues to do so. Mesmerising animals once widespread in the country have declined or become extinct. Javan and Sumatran rhinos (*Rhinoceros sondaicus* and *Dicerorhinus sumatrensis*, respectively), and Sumatran tigers (*Panthera tigris sumatrae*) have decreased to within a hair's breadth of extinction, while Javan tigers (*Panthera tigris sondaica*) died out in the 1970s. Orang-utans (*Pongo* spp.), probably the best-known Indonesian animal, have declined by at least 50 per cent over recent decades.

Environmental degradation on a massive scale underpins Indonesia's wildlife crisis (Petrenko et al., 2016). A major cause of deforestation is the logging industry, driven by demand from China and Japan. About 73 per cent of this destruction is believed to be illegal. Large areas of forest have been cleared by multinational companies and subsequently replaced by plantations. On top of this, woodland is often burnt by farmers and plantation owners. The smoke wafting from fires in the tropical forests of Indonesia form plumes large enough to blot out the sky in Malaysia and Singapore. Around 80 per cent of the fires in Indonesia are set to clear land for palm oil plantations, which substantially reduce species richness compared with primary and secondary forests. All in all, Indonesia has become an ongoing ecological catastrophe. The human population in the country was the fourth largest in the world in 2020, and increasing at about 1 per cent annually. Transmigration of people within Indonesia from the population-dense island of Java to other parts of the country has contributed to forest loss, as migrants often face hostility from incumbent residents. They then have to invade new forest areas to make a living, and there is a positive relationship between migration and deforestation. Population pressure is therefore one of the drivers of the biodiversity crisis in Indonesia.

Of the tropical regions considered here that are suffering from extensive environmental degradation, only Indonesia has a human population density, at 151 people per km^2, comparable with developed countries of the EU (averaging 112 people per km^2). East African states have around 80, the Sahel has 47 and Brazil has just 25 people per km^2. All of these values are far outstripped by the UK, at 275, and most dramatically by England at 432 people per km^2. The developing countries therefore still have relatively small populations compared with most of Europe but are now at the stage that that continent experienced many centuries ago when it too underwent massive habitat destruction. Expansion of humans into pristine tropical habitats certainly follows on from population growth in many cases, but this in turn relates to poverty and lack of alternative income sources in these biologically important regions. Long-term conservation of invaluable and precious landscapes requires both population growth and privation to be addressed.

How does biodiversity loss relate to human population size in the world as a whole? A Biodiversity Intactness Index (BII) (Newbold et al., 2016) was used to attempt quantitative comparisons of wildlife declines across the planet. The BII estimates the average abundance of originally present species across multiple taxa, relative to their percentage abundance in undisturbed habitats. The calculations are complex and vary somewhat according to geographical criteria, but it is possible to list 122 countries around the world with rankings that fall within 15 places on separate mapping systems. BII estimates varied from as low as 62 per cent in the Philippines up to 95 per cent in Greenland. These BII values correlate strongly, and negatively, with human population density, as shown in Figure 9.2 ($r_s = -0.4657$, $P < 0.0001$). Across the world as a whole, more people are associated with more biodiversity loss.

9.3 International Organisations and Population Issues

The consequences of human population increases have not gone unnoticed by international organisations concerned with the welfare of people all over the world. Foremost among these is the UN and

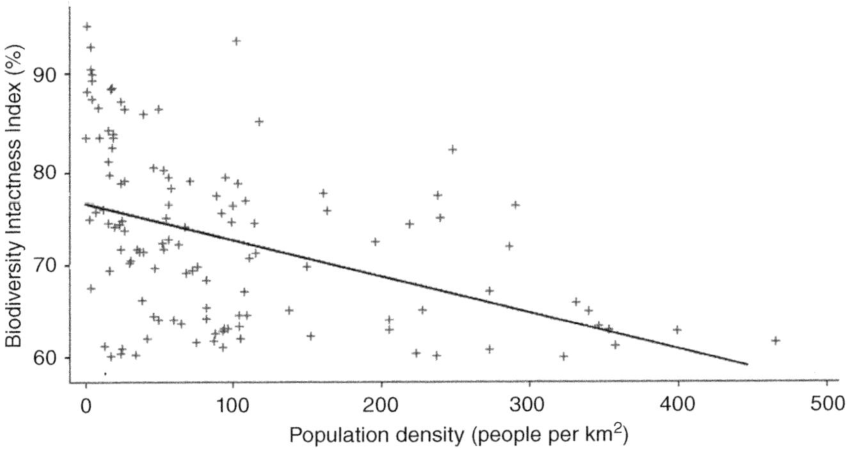

Figure 9.2. Biodiversity intactness and human population density.

Secretary-Generals of this prestigious body have occasionally commented on population matters. U Thant, a Burmese diplomat and the third Secretary-General of the UN from 1961 to 1971, remarked that problems of growing food shortages could not be solved in many cases without simultaneous efforts to moderate population growth. Kofi Annan served as the seventh Secretary-General of the UN between 1997 and 2006, and was a co-recipient of the 2001 Nobel Peace Prize. He held forthright views on the dangers of population growth, insisting that any belief that it guarantees a better life is a myth that only those who sell nappies, prams and the like have any right to believe. He realised that population stabilisation should be a priority for sustainable development, with a strong focus on the empowerment of women and girls. UN Under-Secretary-Generals, notably Maurice Strong and Baroness Valerie Amos, have said much the same things, with the warning: 'Either we reduce the world's population voluntarily or nature will do this for us, but brutally.' The UN currently has an active population division, the UN Population Fund (UNFPA), which conducts demographic research and assists countries in analysing population data and information. It regularly publishes data sets on the world's population and analyses global demographic trends. Thoraya Obaid was Executive Director of the UNFPA from 2001 to 2010. She recognised that it was impossible to tackle the massive challenges of poverty, hunger, disease and environmental destruction without addressing issues of population and reproductive health. UNICEF, the UN Children's Fund (originally the

International Children's Emergency Fund), is another agency in this case empowered to provide humanitarian and developmental aid to children worldwide. Previous director James Grant pronounced that family planning could bring more benefits to more people at less cost than any other single technology available to the human race.

Vast international organisations can seem soulless, however high-flying their intentions. Preoccupancy with meetings and eternal diplomacy has left humour rather hard to find within the UN. However, Kofi Annan's successor, Ban Ki-Moon, had some notorious stabs at it. 'Rugby scrums confuse anyone who doesn't know the game. So do UN debates,' he said. 'And sometimes they can look very similar! In rugby, you lose teeth. In diplomacy, you lose face.' This was one of his better efforts. The UN expresses its concern about the future of humanity by way of 17 Sustainable Development Goals (SDGs), none of which explicitly mention population growth but most of which will be unachievable unless that issue is addressed (Figure 9.3). However, they do include 231 unique indicators of progress towards sustainable development relevant to population matters, including: (1) the proportion of women of reproductive age (aged 15–49 years) who have their need for family planning satisfied with modern methods; (2) adolescent birth rates (aged 10–14 and 15–19 years) per 1,000 women in those age groups; and (3) the number of countries with migration policies that facilitate orderly, safe, regular and responsible migration and mobility of people. The SDGs include environmental ones, notably 'Life below water' (no. 14) and 'Life on Land' (no. 15). 'Life below water' includes 10 target areas: (1) reduce marine pollution; (2) protect and restore ecosystems; (3) reduce ocean acidification; (4) sustainable fishing; (5) conserve coastal and marine areas; (6) end subsidies contributing to overfishing; (7) increase the economic benefits from sustainable use of marine resources; (8) increase scientific knowledge, research and technology for ocean health; (9) support small-scale fishers, and (10) implement and enforce international sea law. 'Life on land' seeks to: (1) protect, restore and promote sustainable use of terrestrial ecosystems; (2) sustainably manage forests; (3) combat desertification; and (4) halt and reverse land degradation and halt biodiversity loss. Achieving these admirable and ambitious goals inevitably requires dedicated and effective conservation measures that would be more likely to succeed if population pressure was reduced.

The role of the UN therefore mostly involves giving advice about sustainability to member states, while providing information on topics including population growth upon which they may or may not act.

Figure 9.3 The UN Sustainable Development Goals. *Source*: UN Sustainable Development Goals, www.un.org/sustainabledevelopment/. The content of this publication has not been approved by the United Nations and does not reflect the views of the United Nations or its officials or Member States (A black and white version of this figure will appear in some formats. For the colour version, please refer to the plate section.)

Contraception is obviously of critical importance to achieving population goals, not least by empowering women to make choices about family size. This ambition has not been helped by the decision of several recent US presidents who, for reasons of religious dogma, have insisted that aid provided to developing countries cannot be used to support contraception programmes or abortion. The United States is a major source of foreign aid, so the adverse consequences for birth control have been substantial, although funding has been substituted in large part by more enlightened voluntary organisations.

The WHO is an international body with the primary goals of ensuring that a billion more people have universal health coverage, protecting a billion more people from health emergencies, and providing a further billion people with better health and well-being. The WHO began its life in April 1948 and now includes over 7,000 people from more than 150 countries, with a headquarters in Geneva. It aims to address issues of non-communicable disease prevention, mental health promotion, climate change in small island developing states, antimicrobial resistance, and elimination and eradication of high-impact communicable diseases. The WHO carries out hugely important projects across the world but does not mention population growth as an issue impacting on its work, let alone have a policy on the subject.

Save the Children (STC) began life at the end of the First World War, as a vision of two sisters appalled at the suffering of starving infants in Berlin and Vienna. Since then, STC has grown dramatically and operates just about everywhere when a need arises. It is made up of 30 member organisations, working in 117 countries. In 2011, STC launched a 5-year 'No Child Born to Die' campaign, which has helped bring about breakthroughs in vaccination, nutrition and newborn health. The goal now, signed by 169 national leaders in 2015, is to end preventable child deaths and provide healthcare and quality education for all children by 2030. STC maintains that there is sufficient food in the world for everyone but that for various reasons not enough people benefit from it. Apparently, a quarter of the world's children in the world today are suffering permanent damage to their bodies and minds because they do not get the nutrition they need, and over 3 million die annually as a result. STC has done wonderful work for over a century, but yet again, there is no recognition that their outstanding achievements are continually compromised by ever-increasing numbers of deserving cases.

9.4 International Conventions and Conferences

Climate Change

Concerns across the world about environmental deterioration have spawned a range of multinational talking shops, variously described as conventions, accords, symposia, conferences and the like. Undoubtedly, the best known of these, and the most successful thus far, was the Paris Agreement on climate change adopted by 196 parties in Paris on 12 December 2015. This legally binding international treaty aimed to limit global warming by dramatically reducing the output of greenhouse gases. The Paris Agreement was a landmark development because it followed a series of unsuccessful attempts to create an international consensus on the need to act on global warming. In 1992, a UN Conference on Environment and Development, the 'Earth Summit', held in Rio de Janeiro, produced the UN Framework Convention on Climate Change (UNFCCC), signed by 154 states. Subsequently, the Kyoto Protocol was signed in 1997 and came into force in 2005. This was the first implementation of measures under the UNFCCC. Compliance with the Kyoto Protocol has been patchy. It put the primary obligation to reduce emissions on developed countries, on the basis that they were historically responsible for the elevated levels of greenhouse gases in the atmosphere. Although 36 major developed countries reduced their emissions, the global output of damaging gases increased by 32 per cent between 1990 and 2010. In 2009, the 15th session of the Conference of the Parties to the UNFCCC took place in Copenhagen. The subsequent Copenhagen Accord contained several widely agreed key elements. These included the long-term goal of limiting the maximum global average temperature increase to no more than 2°C above pre-industrial levels, subject to a review in 2015. There was, however, no agreement on how to do this in practical terms. Fortunately, the realisation that climate change is not only real but poses a significant, real-time threat to people everywhere finally broke through previous inertia at the Paris summit. Although action needs to increase massively to achieve the goals of the summit, the years since its agreement have already sparked low-carbon solutions and new markets. Increasing numbers of countries, regions, cities and companies are establishing carbon neutrality targets, and zero-carbon solutions are becoming economically competitive. For many naturalists there is a feeling of déjà vu about all this, as they were among the first to recognise the effects of climate change acting on wildlife phenology and distributions (Beebee, 2018).

This was all very promising, but what about the contribution of continued population growth to climate change? This topic is not explicit in the Paris Agreement, but its relevance as an important factor is widely recognised. A letter warning of a climate emergency, signed by more than 11,000 scientists, called for a gradual reduction in the world's population (Ripple et al., 2020). The signatories listed both economic growth and a global population increase as among the most important drivers of increases in CO_2 emissions from fossil-fuel combustion. The report called for bold and drastic transformations regarding economic and population policies, focused on proven and effective strategies that strengthen human rights, while lowering fertility rates and lessening the impacts of population growth on greenhouse-gas emissions and biodiversity loss. Former US Vice President Al Gore was foremost among politicians recognising the perils of climate change, and also in warning that overpopulation fosters global warming. A study by the Universities of Lund and British Columbia suggested that the single most effective measure an individual in the developed world could take to cut their carbon emissions over the long term would be to have one fewer child. This calculation relied on estimates of future per-capita climate emissions, which are likely to change significantly, so it must be treated with caution. Nevertheless, figures produced by the authors suggested that this strategy could be over six times more effective than other routinely recommended behavioural changes (including going car-free, avoiding one transatlantic flight, choosing a plant-based diet and recycling) put together, in reducing carbon emissions. Evidently, restraining personal reproduction is a medium- to long-term method for reducing climate change, and for now the other, well-publicised methods mentioned above will have to suffice. We are inveigled into adopting more vegetarian or vegan diets, and into travelling less frequently than in the past, whether or not we find these lifestyle choices attractive. This is one of many prices to pay for an ongoing failure to control the human population at a lower, sustainable level.

Wildlife Conferences

The Earth Summit in 1992 was not just about climate change. Other issues were up for discussion, including global wildlife crises. One follow-up was the Convention on Biological Diversity (CBD), which opened for signature in Rio de Janeiro on 5 June 1992. The Convention recognised for the first time in international law that the conservation of biodiversity is a concern of all humankind. The agreement covers all ecosystems and species. The 10th Conference of the Parties to the CBD, in October 2010, adopted a Strategic Plan for Biodiversity 2011–2020, along with 20 Aichi Targets. These targets sat under five strategic goals aimed at stimulating effective and urgent action to halt the loss of biodiversity in signatory nation states. The Convention required that signatory countries prepare national biodiversity strategies. As early as 2012, 173 parties had developed such National Biodiversity Strategies and Action Plans as the principal instruments for implementing the Convention at national levels. Such ambitions were intended to ensure that, by 2020, ecosystems would be resilient, thereby securing the planet's variety of life. By 2016, the Convention had 196 parties, including 195 countries and the EU. The UK prepared elaborate responses to conserve species and specific habitats in the UK, and the Joint Nature Conservation Committee (JNCC) helped to produce the country's 6th National Report to the CBD, published in 2019. This report focused on work to achieve the 20 Aichi Targets in Britain. With respect to funding for international biodiversity conservation, the UK's contribution increased from around £77 million per year between 2006 and 2010, to over £180 million in 2015. This is remarkable because it covers a period during which funding for statutory nature conservation within the UK virtually halved. Not surprising, then, that the national picture in Britain is one of ongoing species decline. Despite some progress in improving the condition of protected sites, a significant proportion of the best wildlife habitats inside and outside these sites have remained unfavourable.

It is not only in Britain that the CBD plan has fallen short. In its 2020 report, a pessimistic tone made it clear that none of the 20 targets were fully met, and that just six were partially achieved. This was hardly a surprise. A mid-term assessment in 2014 already suggested that the targets were off track. WWF's 2020 Living Planet Report estimated that, globally, populations of nearly 21,000 species of mammals, fish, birds, reptiles and amphibians plummeted by an average of 68 per cent between 1970 and 2016. All of the ambitious CBD projects took virtually no account of how increasing human numbers were compromising their conservation plans. Only one lone voice pointed out that 'with increasing human population and increasing development and consumption levels, nature will decline in coming decades – there's no question about that'. Rather a shame that such an important problem for the future of biodiversity was so dismissively glossed over.

Then there is the Intergovernmental Science-Policy Platform on Biodiversity and Ecosystem Services (IPBES), an independent intergovernmental body established 'to strengthen the science-policy interface for biodiversity and ecosystem services'. It was established in Panama on 21 April 2012 by 94 governments, intended to serve a similar advisory role to the Intergovernmental Panel on Climate Change. An IPBES Global Assessment Report in spring 2019 painted a damning picture of the state of the world's ecosystems (https://pfbc-cbfp.org/news-partner/Assessment-Report.html). Compiled by 145 expert authors from 50 countries with inputs from another 310, and based on a systematic review of about 15,000 scientific and government sources, it concluded that the health of ecosystems on which we and all other species depend is deteriorating more rapidly than ever. The report found that around 1 million animal and plant species are threatened with extinction, many within decades, and the history of extinctions has progressed continuously as a result of human activities (Figure 9.4). The average abundance of native species in most major land-based habitats has fallen by at least 20 per cent since 1900. More than 40 per cent of amphibian species, almost 33 per cent of reef-forming corals and more than a third of all marine mammals are threatened. At least 680 vertebrates have been driven to extinction since the sixteenth century. Five main drivers of change were identified: (1) changes in land and sea use, (2) direct exploitation of organisms, (3) climate change, (4) pollution, and (5)

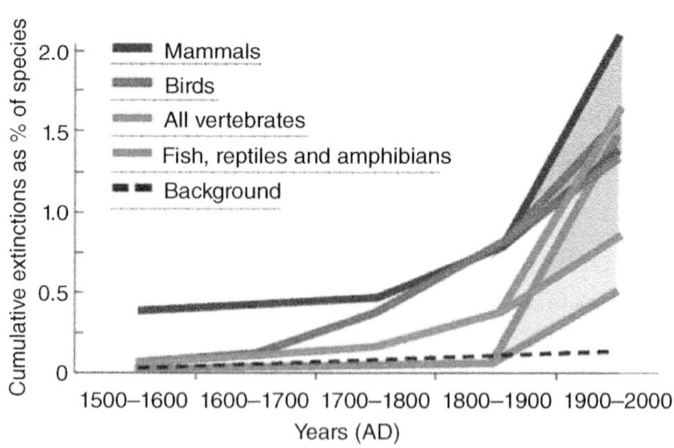

Figure 9.4. Patterns of animal extinctions over time. Permission courtesy of British Geological Survey © UKRI. *Source*: https://science.sciencemag.org/content/351/6269/AAD2622.abstract (A black and white version of this figure will appear in some formats. For the colour version, please refer to the plate section.)

invasive alien species. As usual, there was barely any mention of underlying population pressure, with just one off-hand remark that since 1970 the global human population has more than doubled, from 3.7 to 7.6 billion. There was no suggestion that remedial action to recover a rapidly disappearing biodiversity might include policies on this matter.

Population Conferences

Climate change and biodiversity crises have dominated environmental concerns for decades, but worries about population growth have not been entirely ignored. An initial World Population Conference was organised in Rome in 1954 to exchange scientific information on demographic variables, their determinants and their consequences. This was the start of a series of such meetings. The second one was held in Belgrade in 1965 and emphasised the importance of fertility analysis as part of development planning policies. In 1974, the Third World Population Conference in Bucharest led to production of the *World Population Plan of Action* (UN Population Information Network, 1974). This extraordinarily wordy treatise deters all but the most dedicated of readers to peruse its wide-ranging and frequently repetitive paragraphs. It emerged at a time when concern about population growth was high on the international agenda, yet managed to avoid making serious recommendations on how to deal with it. The plan maintained that the essential objective was social, economic and cultural development of countries, and that demographic change and development were interdependent. Some of its claims included that 'Of all things in the world, people are the most precious. Man's knowledge and ability to master himself and his environment will continue to grow. Mankind's future can be made infinitely bright.' It tried to be even-handed in outlook, suggesting that countries aiming at moderate or low population growth should try to achieve it through reduced birth and death rates, while those wishing to increase population growth should, where appropriate, encourage increases in fertility and immigration. Regardless of overall demographic goals, it emphasised the right of persons to determine, in a free, informed and responsible manner, the number and spacing of their children. Some social welfare programmes, such as family allowances and maternity benefits, that may have a positive effect on fertility may be strengthened when such an effect is desired. These programmes should not be curtailed if the opposite effect on fertility is wanted. It is hard to glean any serious concern about overpopulation from its vast and convoluted text. The fourth conference of 1984 in Mexico City reviewed and approved most aspects of the Bucharest

Conference Agreements. Human rights, conditions of health and well-being, employment and education were the main issues highlighted in a declaration signed at this conference. The Fifth International Conference on Population and Development (ICPD) was held in 1994 in Cairo. This is the best-known meeting of the series, during which, a new Programme of Action was adopted as a guide for national and international objectives in the field of population and development to last for 20 years. The new agenda re-emphasised the relationship between population and development, and focused on the needs of individuals rather than simply responding to demographic goals.

For many observers hoping for clear policies to constrain population growth, the outcome of the Cairo meeting was disappointing. It was evident from the start that the draft Programme of Action to be discussed at this UN International Conference was not about population and development but about women and related agendas. It placed reproductive health and rights, women's empowerment and gender equality at the heart of the development agenda but did little to link those goals with global population growth. Feminists claimed a major victory at the Cairo summit, insisting that slowing population growth must henceforth be addressed within the larger context of human rights for women. Unfortunately, crucial discussions of economic development were neglected at the 10-day meeting and environmental concerns were almost completely ignored (Figure 9.5). All the issues about women's rights discussed in Cairo are vitally important, but it is surely regrettable and unnecessary that they distracted so effectively from overpopulation issues.

Two subsequent sessions of the UN General Assembly were held in 1999 and 2014 to assess implementation of the Programme of Action adopted at the 1994 conference, but the most important review came in 2019, at the Conference on Population and Development in Nairobi. This summit aimed to be maximally inclusive and brought together heads of state, ministers, parliamentarians, thought leaders, technical experts, civil society organisations, grassroots societies, young people, business and community leaders, faith-based organisations, indigenous peoples, international financial institutions, people with disabilities, academics and many others interested in the pursuit of sexual and reproductive health and rights. A majority of the 170 countries represented signed up to 1,250 concrete commitments for increasing budgets for health, enabling access to modern contraceptives and for training more midwives, as well as vaguer promises such as 'harnessing the demographic dividend' from their young populations. But the Conference lacked a

Figure 9.5 The Cairo population conference of 1994. *Source*: UNFPA, www.unfpa.org/news/explainer-what-icpd-and-why-does-it-matter

sense of perspective as to the impacts of global human population growth on people and planet. The Nairobi summit was very much focused on pursuing the Programme of Action agreed at Cairo 25 years earlier, enshrining an attitude that has held sway ever since. No longer was attention to be directed towards the impacts of population growth, and certainly not towards any attempt to reduce this growth. Instead, efforts should proceed exclusively on improving reproductive health and women's rights. Tackling these issues, especially addressing unmet needs of young women and girls, was both right and necessary. But effectively maintaining a taboo around putting these important matters in the context of ongoing population growth was a mistake. Following this shift of emphasis at Cairo, the global population increased by over 2 billion from 5.6 billion in 1994 to 7.7 billion in 2020. Despite the focus on enabling more women in developing countries to access safe, modern contraception, the proportion of women doing so increased by only 6 per cent globally, from 52 to 58 per cent. Some 232 million women worldwide still have an unmet need for modern contraceptives. Consequently, the progress being made in slowing and stabilising population growth pre-Cairo has stalled and even reversed. For all the good intentions, human numbers have continued to grow by an additional 80 million people per year. Closing off any overt discussion about the human population size was evident from the start in the agenda set for the Nairobi summit, producing yet another missed opportunity.

How has the UK government responded to population concerns? For World Population Day 2018, Population Matters called for government action in response to the second *World Scientists' Warning*, published in 2017 and signed by more than 20,000 scientists (Ripple et al., 2017). The warning spoke bluntly about the risk of widespread misery and catastrophic biodiversity loss unless urgent action was taken to address environmental problems. Those actions included boosting support for girls' education and family planning in order to reduce family size and population growth, and estimating a scientifically defensible, sustainable human population. The government's response came 3 weeks after the call. Rather than addressing the full range of environmental threats identified in the warnings, it concentrated on just one – climate change. Far from recognising the need for the fundamental changes the *World Scientists' Warning* called for, it was a generic, essentially noncommittal reply. Population issues remained firmly off the agenda.

9.5 Overview

The continuing expansion of human numbers has had far-reaching effects across the world. Wildlife has been a major casualty of this trend, not just in the UK but also in many other countries. Some of these, particularly in the tropics, are the main reservoirs of global biodiversity. There are many pressures arising from unfair distribution of resources favouring the developed world, but population growth in the developing world is indisputably one widespread and significant driver of environmental degradation.

Outspoken individuals within major international bodies concerned with human welfare, such as the UN, have remonstrated about the dangers of ongoing population growth. Unfortunately, their voices have had limited impact. Official UN responses have addressed environmental issues with 17 SDGs, but promising though they are, this approach has skirted around any concept of a full-blown population policy. Conferences are potentially powerful fora for achieving international consensus. The 2015 Paris Agreement on climate change is an example of how this can work effectively, and previous international assemblies successfully generated international action on crises including acid-rain pollutants, and on chlorofluorocarbon gas containment to protect the atmospheric ozone layer. Wildlife conferences have been less valuable, mostly involving wringing hands about ongoing declines but broadly failing to arrest them. Population summits have arguably been the least useful of all, diverted from direct action on policies for limiting population growth into ancillary aspects often peripheral to the central issue. An underlying attitude seems to be that generally declining birth rates will sort out human population growth problems without the need for specific actions. This ignores the fact that human numbers are already too high for long-term sustainability, and that extra billions of people are still expected to arrive during the twenty-first century even in the most optimistic scenarios. This is surely a failure of international responsibilities by a wide range of governments and international organisations. It amounts to a spineless response, shying away from a problem already serious and set to get worse. The consequences will haunt us for decades, if not centuries to come.

10 · *Conservation in a Crowded Country*

10.1 Approaches to Wildlife Conservation in Britain

Protecting wildlife is a difficult ambition at the best of times, not least because conflicts of interest are almost universal in conservation work. Whether in the game parks of East Africa or on the sand dunes of England, competing pressures are everywhere. Inevitably, increases in human numbers exacerbate such difficulties. As discussed in Chapter 1, the UK, and England in particular, has among the highest human population densities in the world. Britain is home to a total of over 3,800 vascular plant and 27,000 insect species, while Scotland has around 1,600 plants and 14,000 insects. Although much of this diversity is shared between England and Scotland, just over 25 per cent of the UK's 969 wildflowers listed by Sterry (2006) do not occur north of the border. By contrast, less than 4 per cent are confined to Scotland. England's greater biodiversity is consequent on its warmer southerly position as well as a more varied geology than Scotland, but this leaves a majority of British species struggling to survive in the most densely populated country of the union.

To cope with the challenges of wildlife conservation, a hierarchy of methods designed to protect wild plants and animals has been assembled, and is broadly congruent across the world. Top of the inventory is habitat protection with manifestations including National Parks, nature reserves and other designated areas. Then there is species protection, disallowing collection, sale or killing of listed organisms. And last but certainly not least is prescriptive management to maintain habitats and their inhabitants in situations where just leaving them alone does not suffice.

Landscape-level Protection

The relative importance of the various wildlife conservation measures varies considerably according to local circumstances in specific countries, but habitat safeguarding is invariably vital for long-term success. National Parks are the largest areas granted protection from development, but their value for wildlife depends very much on national priorities. Large countries have the largest parks. The USA has 60 of them; the largest (in Alaska) covers 32,000 km^2 and is twice the size of Yorkshire, England's largest county. All of the US parks put together account for more than 210,000 km^2, which is about 2.3 per cent of the total US land area. The UK has about 22,000 km^2 of National Parks, with the Cairngorms in Scotland comfortably the largest at 4,500 km^2. Britain's 15 National Parks, mostly much smaller than the American ones, cover 7 per cent of the land area in Scotland, 9 per cent in England and almost 20 per cent in Wales. This higher percentage of national park land cover in the UK relative to the USA is not as good as it looks from the perspective of wildlife conservation. In the USA, this is the primary objective of National Parks, and although tourism is encouraged, there is very little other economic activity. Scanning Yellowstone's lush valleys with their vast herds of grazing bison, surprising a bear family while on an afternoon ramble or stumbling on a hunting wolf pack, anyone visiting these vivifying landscapes is left in no doubt about why they exist.

The opposite is true for most parks in the UK where tourism is also promoted but very much within a working countryside. Farming is extensive and commonplace, and wildlife conservation is low on the agenda. National Parks were created with public access and enjoyment very much in mind, but Glover (2019) produced a damning critique of the health and management of the country's National Parks. He urged that they be made 'greener, more beautiful and open to everyone' and to 'reignite the founding spirit in which they were created'. If wildlife can thrive anywhere, it should be in Britain's National Parks. Yet nearly 75 per cent of the SSSIs in England's National Parks are in an 'unfavourable' condition, and the total UK figure is little better, at 61 per cent. The IUCN places the UK's National Parks in category V, which is the second lowest out of six. Ongoing pressure for greater access has been detrimental to the landscapes that visitors flock to see. Hundreds of people have protested in the Lake District against proposals to allow holiday house boats on lakes, introduce a zip-wire and continue to allow off-road driving. Debbie Tann of the Hampshire & Isle of Wight Wildlife Trust emphasised that an effective national park needs to focus more on nature recovery than on recreation. She points to the New Forest, a small national park surrounded by large population centres: 'From a recreational point of view the national park is great but its biodiversity is in decline.' Ground-nesting birds such as nightjars and woodlarks are disturbed by dogs, while soil erosion and compaction is being reported both on road verges and along the busier footpaths. The New Forest is under pressure because housing developments have failed to take account of a need for more open spaces for their new inhabitants. This means that people naturally turn to the National Park, expecting that it can absorb indefinite increases in public pressure. The forest is constantly nibbled away at the edges and the place is simply unable to cope with a large footfall. Arguably, the National Parks are in an impossible position as, despite their nominal protection, they too often cannot serve wildlife effectively in the face of burgeoning public pressure. AONBs, the poor relations of National Parks, share similar management approaches and were created to celebrate some of the country's most attractive scenery. England, Wales and Northern Ireland have 46 AONBs, covering 18 per cent of the countryside. That of the Mendip Hills, adjacent to our home village, is mostly indistinguishable from the surrounding Somerset countryside in terms of land use, although the administration does have a remit that includes some wildlife conservation work (Figure 10.1).

In 2020, the government announced plans to increase landscape-level protection, including the creation of new National Parks and AONBs, as

Figure 10.1. Cheddar Gorge in the Mendip Hills, an AONB. *Source*: Matt Cardy/Stringer/Getty Images

part of its aim to protect 30 per cent of the UK's land by 2030. There are proposals to amalgamate National Parks and AONBs into a category of 'Natural Landscapes' with a renewed mission to recover and enhance nature. Apparently, by 'designating more of our beautiful and iconic landscapes as National Parks and AONBs, and through the new Landscape Recovery projects, we will help expand and protect precious wildlife habitats and, vitally, increase people's access to our treasured landscapes' (GOV.UK, 2020).

Wildlife protection is surely, in current circumstances, the more vital ingredient and it is hard to see how adding more sites with escalating public pressure will do much to ease Britain's wildlife crisis.

Then there is a range of local protection measures as described in Chapter 4, with SSSIs as the most widespread designation specifically to protect wildlife. As also mentioned earlier, SSSIs are sacrosanct – until they aren't. Time and time again, these most precious jewels in the wildlife crown succumb to challenges from other vested interests, including housing development, road construction and ad hoc damage by farmers. Landowners of SSSIs are responsible not just for their physical protection but also for carrying out prescribed management to maintain their ecological value. In 2020, less than 40 per cent of SSSIs in England were judged to be in favourable condition. Much of the deterioration identified by surveyors was down to neglect, exacerbated by underfunding of essential works and failure of the statutory authorities to enforce legal responsibilities. So much for the best landscape-level approach allegedly safeguarding the UK's wildlife heritage. It is not hard to make a connection between this sad situation and the pressures of an increasing human population outlined in previous chapters.

Species Protection

Legislation in a series of Parliamentary Bills passed over more than half a century ago has generated a long list of species for which persecution in any form, often as well as disturbance or damage to their habitats, is illegal. These measures are often targeted at activities that amount to some form of predation by humans, but in many cases, this is not an important cause of species declines. Catching a few great crested newts, or even natterjack toads, is against the law unless carried out under licence, but it is habitat loss or change that has done for these animals. Arguably, this type of heavy-handed, Big Brother approach distances people, especially children, from engagement with wildlife and can

Figure 10.2. A legally protected pine marten. *Source*: Horst Helwig/McPhoto/ullsteinbild/Getty Images

therefore be counterproductive. That is not to say that directing protection at particular species is always pointless. It comes into its own when rare species subject to persecution are involved. Raptors and mammalian predators such as pine martens (Figure 10.2), polecats and wildcats have clearly benefitted from protective legislation. Even so, the relevant laws have not always worked well. The flouting of raptor preservation laws on shooting estates has remained resilient to full prohibition, and destruction of ponds occupied by great crested newts has gone on regardless, with at best the adoption of feeble mitigation methods. Most species of wild birds have enjoyed legal protection in the UK since the 1950s, yet many of them have declined seriously since that time. There was a time in the early post-war period when laws to protect individual or groups of species were considered an adequate response to conservation concerns, but it rapidly became clear that such measures alone were insufficient. The real pressure was and remains destruction and mismanagement of habitats.

Habitat Protection and Management

Most of the British countryside lies outside any kind of protected area, and it is in this inherently vulnerable zone that most wildlife has to survive. Even within nominally protected places such as SSSIs, intrusive management is often required to sustain viable ecosystems. At one level, habitat protection is simply defending areas against development threats and, as exemplified in earlier chapters, this has often failed. These threats are many and varied, and sometimes unexpected. Measures to counter climate change have included the construction of wind farms that can kill birds and bats in considerable numbers. Tidal barrages, thus far largely figments of developers' imaginations, would be likely to have hugely negative impacts on saltmarsh communities and migratory fish. It is in the management of farmland, however, that the most damaging changes have occurred. As outlined in Chapter 5, agricultural intensification has been the single largest driver of wildlife declines. Efforts at conservation here have involved, over recent decades, dedicated areas of set-aside land where habitat is allowed to regenerate, leaving wide field margins unploughed so wild plants and pollinators can thrive, and countryside stewardship schemes where farmers are paid to carry out wildlife-friendly management. These schemes, which can include hedge extension, copse creation and pond restoration, have proved the most beneficial interventions, but they inevitably involve only a small proportion of the farmed landscape. Elsewhere, intensive operations including the use of pesticides continue apace. (Re)wilding, as is occurring at Knepp in Sussex (Tree, 2018), offers the prospect of agricultural deintensification on a much larger scale, but in England at least this is unrealistic other than in a few places. Land restored in this way usually relies heavily on government subsidies rather than food production to be economically viable, and in a densely populated country with increasing concerns about food security, it is not a formula likely to be widely adopted. It may have a brighter future in Scotland and Wales (I hope so), where more suitable space is available, but in these countries wildlife is less diverse and not under as much threat as it is in lowland England. Linking areas of the ancient Caledonian Forest westwards from the Abernethy RSPB reserve in the Cairngorms is a promising rewilding operation that should benefit many precious and rare species. But across too much of the English landscape, wildlife only flourishes in habitat islands, widely separated by inhospitable terrain. This problem was recognised by the government (Lawton et al., 2010) and requires mechanisms to increase connectivity across the countryside, which, to be realised, will require extensive investment on farms to further extend hedgerows and create more woodland and ponds, all on a larger scale than stewardship schemes have been able to achieve.

Departure from the EU has precipitated another change for farmland economics, and potentially also for wildlife conservation. The generous subsidies previously provided to sustain cheap food production at whatever cost to the environment are being phased out over several years, to be replaced by yet another programme. The new 'environmental land management scheme' (ELMS) will fund environmental improvement efforts such as enhancing soil health, creating natural flood barriers, species management, and restoring woodland and peatland. The slogan for ELMS is 'Public money for public goods', aimed at disease eradication, nature protection, high air and water quality, more tree planting and greater biodiversity. All of this, by moderating the impact of intensive agriculture, sounds promising, but the road to arrest wildlife declines has been paved with good intentions. Like its predecessors, ELMS will face the continuing challenge of maintaining adequate food production with a better future for wildlife. With the human population still increasing, this looks like a highly optimistic ambition.

Habitat Management in Practice

In the face of unrelenting pressure on wildlife in the British countryside, it is the generally small-scale activities of NGOs that have proved the most important in slowing the losses of biodiversity in the UK. The country is astonishingly well provisioned with charitable organisations employing professional staff and supported by huge numbers of volunteers, all committed to maintaining wildlife in so far as they are able. Wildlife NGOs range in size from those employing hundreds of people with nature reserves all over the country, of which the RSPB is a prime example, to a multitude of smaller concerns specialising in particular species groups of Britain's flora and fauna. As one example, the Amphibian and Reptile Conservation Trust is dedicated to the effective protection of frogs, toads, newts, lizards and snakes, several of which are endangered. As a trustee, I have witnessed at first hand the dedication and enthusiasm of the staff and volunteers, and the successes they have had in arresting declines of these animals (Figure 10.3). It is a story mirrored by conservation NGOs all across the land. Scrub bashing, pond restoration and creation, tree planting, propagation of wildflower meadows and much more have been the mainstay of effective conservation for decades. Some efforts, especially those in which different groups collaborate, have made dramatic contributions to wildlife conservation. A conspiracy of the RSPB, the Somerset Wildlife Trust and Natural England created the Avalon Marshes on the

Figure 10.3. Amphibian and Reptile Conservation (ARC) volunteers carrying out heathland management work in Surrey. © Bryony Davison, ARC (A black and white version of this figure will appear in some formats. For the colour version, please refer to the plate section.)

Somerset Levels. In a landscape previously ravaged by uncontrolled peat extraction, this extensive wetland is home to a wondrous collection of rare birds, invertebrates and plants attracting visitors from far and wide. Bitterns and cranes are thriving again, but these success stories must be viewed in a broader perspective. The bigger picture, even on the Levels, is of increasing eutrophication and poorer water quality in the region's rivers, and ongoing declines of species outside the conservation zones. Large areas of the Somerset wetlands now enjoy statutory protection, and rich wildlife abounds across much of the region, but nightingales rarely give voice and whinchats (*Saxicola rubetra*) have disappeared as a breeding species. Here as elsewhere, conservation efforts have successes to celebrate, but in the round, they are rearguard actions that have not arrested the overall pattern of wildlife declines in the UK.

10.2 Population Policies around the World

The UK is just one country with wildlife declines related to human overcrowding. The BII described in Chapter 9 shows how widespread this problem is. Below are some examples of how biodiversity declines relate to population policies in countries around the world.

Well-known Examples of Population Policies

Some countries have made unilateral efforts to contain population growth since the mid-twentieth century. Two of these are particularly notorious. As mentioned in Chapter 8, India initiated a program of compulsory mass sterilisation for men with two or more children in the 1970s. This is surely an archetypal example of what not to do. The inevitable backlash set back population debate in India for decades, with people numbers increasing more than twofold from 600 million in the early 1970s to almost 1.4 billion in 2020. Better news is that, despite the antipathy that followed the sterilisation scandal, India's fertility rate has been in continuous decline since the mid-1980s, down to around 2.1 children per woman in 2020. Nevertheless, with such a high baseline, an extra 100 million people are still expected by 2036, by which time India will have overtaken China as the most populous country on Earth. India is already in serious trouble because of diminishing resources. An increasing population and inadequate surface water is fast depleting the country of its groundwater reserves. It is one of 17 countries where water stress is extremely high, according to a recent global report. Currently a major food producer, India's agricultural output is threatened by the deteriorating water situation. Farmers report that, over the years, borewells have had to be made ever deeper. Initially dug down to 45 m to find water, 10 years ago they had to double that depth. Now it is necessary to go down to 153 m. There are serious concerns that soon there will not even be enough drinking water, let alone water for agriculture. This is not a good situation with many more millions of mouths to feed anticipated in the near future. India has a BII of 65.5 per cent, indicating a substantial biodiversity deficit on top of the impact of overpopulation on human lives.

Without doubt, the best-known attempt to curtail population growth was China's one-child policy. Introduced in 1978, the programme was enforced in 1979 after a decade-long two-child policy. It was subsequently modified in the mid-1980s, allowing rural parents a second child if the first was a daughter, and in 2015 came a reversion to a two-child limit. In 2021, this was relaxed further to encourage three-child families. The effects of the one-child policy are controversial, with the Chinese government claiming that 400 million births were averted after 1970, thus including the preceding period with a two-child limit. The social consequences of this draconian population policy were extensive and often appalling. Enforcement included a 'family planning fine'. Women were required to have a contraceptive intrauterine device (IUD) surgically installed after having a first child, and to be sterilised by tubal ligation after having a second one. From 1980 to 2014, 324 million Chinese women were fitted with IUDs in this way and 108 million were sterilised. Those refusing these procedures could lose their government employment, their children would be debarred from access to education or health services, and any pre-existing privileges might be revoked. For parents who had 'unauthorised' births, or who wanted a son but had a daughter, giving up a child for adoption was an acceptable strategy to avoid penalties under the one-child restrictions. Many such children ended up with foreign parents far away from China, and, even worse, some baby girls simply 'disappeared'. The cultural preference for male children also caused a substantial imbalance in the sex ratio, leading to too few potential partners for the extra men. Unlike the situation in India where politicians instigating unpopular control measures were at least democratically accountable, no such safeguards protected citizens of the Peoples' Republic of China. Mao Zedong's government was not constrained by humane reproduction policies in devising methods to reduce China's population. In the Cultural Revolution of the late 1960s, an inadvertent contribution to population control occurred when tens of millions of the country's citizens died from starvation due to the agricultural reforms termed the Great Leap Forward.

Ironically, some politicians are bemoaning China's recently reduced fertility rate based on the same tired economic arguments described in Chapter 8. On current trends, its human population is set to stabilise within a decade and then fall substantially for the rest of the twenty-first century. China has 16 of the world's most polluted cities, and more than 400,000 people die each year as a result of air pollution. An estimated 190 million people drink water so contaminated that it makes them sick, and some 40 million have had to migrate because their local ecology can no longer sustain them. A particularly serious challenge is a lack of usable water. Skyrocketing demand, population pressures, inefficiencies, overuse and radically unequal geographical distribution combine to leave two-thirds of China's cities without enough water for their needs and 100 of them facing severe shortages. Some 40 per cent of China's land is affected by soil erosion, and the Chinese desert, already a quarter of its land surface, is expanding at a rate of over 3,000 km^2 per year, and is encroaching on Beijing. This is surely a country that should be celebrating the prospect of reduced population pressure, not attempting to counter it for spurious economic reasons. With a BII of 63.1 per cent, there is again an existing biodiversity deficit that attempts to stimulate population growth are unlikely to improve.

Given the damage already underway around the world consequent upon having too many people around, it is perplexing to note that not all nation states are on board with the need to reduce their population sizes – far from it. In the developing world, by 2015, 50 per cent of governments with policies on the subject were making plans to reduce population growth, but in stark contrast, 45 per cent of administrations in developed countries were incentivising higher birth rates. Why the difference? The problems of increasing numbers in poor countries are those discussed in earlier chapters, notably the generation of increased poverty and environmental damage. The perceptions in richer, developed nations have been more equivocal. Panic has sometimes followed declining birth rates, primarily due to concerns about supporting ageing populations but also, seemingly, for nationalistic, essentially tribal reasons. Population size relative to other states can be considered an important demonstration of strength and security. One major power, the USA, does not have an internal population policy but has regularly attempted to influence those elsewhere by making aid to developing countries conditional on compliance with US values on abortion and contraception. In 2010, a Department of State position paper stated that 'The U.S. does not endorse population stabilisation or control' and that 'the "ideal" family size should be determined by the desires of couples, not governments' (US Department of State, 2010). Not everyone agrees. Lindsay Grant appealed to the Obama administration, claiming that the country had wasted 38 years of good lead time in which were added 104 million new Americans who must find food, shelter and employment. The appeal went nowhere. The main population issue for the USA in recent times has centred around immigration, particularly along the Mexican border. Recent allegations of racism in the treatment of Latin Americans attempting to enter the USA has distracted from considerations about overall numbers of people in the country. The USA has a BII of 63.4 per cent, leaving no room for complacency with respect to wildlife conservation as the country's human population continues to grow.

Australia's attitude to population issues is broadly similar to that of the USA. The government does not specify a target population level or rate of population growth. The country's fertility rate fluctuated between 1.7 and 2.0 births per woman over 42 recent years, but, along with nations such as New Zealand and Israel, Australia has seen some of the highest population growth rates in the developed world. Much of this has been fuelled by immigration. In 2017–2018, net overseas migration was 236,700 while the 'natural' increase was 153,800. In total, Australia's population grew by 390,500, or 1.6 per cent. Immigration has been generally welcomed, giving perceived economic benefits priority over increasing congestion in major cities. Even so, the history of immigration has not always been a happy one. In the immediate post-war period, an assisted migration scheme welcomed new arrivals provided they were white. This policy was finally abandoned in the early 1970s. More recently, Australian immigration legislation was changed dramatically with mandatory detention of unauthorised arrivals, coming mostly from Asia and popularly referred to as 'boat people'. A 2018 poll found that a majority of Australians opposed the current rate of immigration to Australia, with 54 per cent of people saying that the total number of migrants is too high. Australia has a BII of 67.9 per cent, somewhat higher than America but nevertheless worryingly low for such a sparsely populated country.

Policies to Increase Population Size

In Russia, a one-off payment equivalent of £5,800 to families with two or more children was launched in 2007. President Putin opined that 'Russia's fate and its historic prospects depend on how many of us there are . . . it depends on how many children are born in Russian families.' Hitler was of a similar mind and instigated 'Mother's Cross' medals in 1939, with honour levels increasing as a function of how many children women bore. This idea was not entirely new. The Médaille de la Famille française, established in France in 1920, remains in force as a tribute to mothers who raise several children 'in an appropriate way'. This pro-natalist policy encourages three-children families by a cash incentive equivalent to £675 monthly for a mother to stay off work for 1 year following the birth of her third child; by providing a 'large-family card' offering various benefits; by giving reductions on train fares; by income tax rates based on paying less tax, the more children you have; by offering 3 years of paid parental leave, which can be used by mothers or fathers; and by providing subsidised daycare for children under the age of 3 years, and full-time school places for those over 3 years, all paid for by the government. This package has had some effect, as the fertility rate in France (at 1.9 children per woman) is one of Europe's highest but still lower than the replacement rate of 2.1. Germany is moving along a similar route. Almost 800,000 children were born in Germany in 2016, up 7 per cent on the year before. The fertility rate was also up, at 1.59, an increase from 1.5 in the previous year and the highest in more than 40 years. This was generated by a host of family policies to tackle the perceived 'demographic deficit', including an elevated parental leave allowance of up to two-thirds of income for the first year.

Despite continuously increasing human numbers in the world as a whole, several countries have followed in the footsteps of France and Germany. Japan, with a fertility rate of 1.4 children per woman, has eased its previously restrictive rules on immigration to ensure maintenance of an adequate work force. In 2015, the Italian government launched a programme offering €800 per couple per birth to try to boost fertility rates and Greece also instituted 'baby bonuses' for each child. Iran experienced a population boom after the Islamic Revolution in 1979 but went on to implement an effective population control policy. However, this was later perceived to be too effective when the country's annual population growth dropped below 1 per cent. Now vasectomies can no longer be carried out at state-run medical centres and contraceptives are only offered to women whose health might be at risk from pregnancy. Brazil, too, is worried about projections that its population will decline from about 211 million in 2017 to less than 164 million by 2100. Television soap operas portraying small families have been blamed for reduced birth rates in the predominantly Catholic country. Even so, Brazil is at least trying to stem high teenage pregnancy rates based on a campaign urging 'Adolescence first, pregnancy after'.

Countries with right-wing populist governments have also taken a lead in promoting higher fertility, often in combination with anti-migration and 'family values' policies. Poland's health ministry created a video to stimulate procreation, suggesting that 'If you ever want to be a parent, take us rabbits as an example. I know what I am talking about. Our father has 63 children.' Abortion laws are also being tightened, generating large-scale protests. In April 2016, the Polish government launched 'Family 500+' to boost birth rates and reduce child poverty by improving living conditions of large families. Initial results suggested that the programme increased the number of births, up by 13–15 per cent in December 2016 and January 2017 compared with the same period in previous years. However, this brief spike in birth rate was not sustained. In Hungary, *in vitro* fertilisation clinics offer free treatment cycles for all women who want them, as long as they are under 40 and not a lesbian (!), and there is a promise to offer a lifetime exemption from income tax for mothers with three children. The jury is still out on whether these policies work. The fertility rate has risen from 1.23 to 1.48, which is still well below the replacement level, although it is hard to gauge how much of the changes are down to specific policies. In general, attempts by government to increase birth rates even with financial incentives have had little long-term success. People, it seems, make their child-bearing decisions mostly on the basis of personal aspirations rather than on government directives.

Few of these countries have BIIs high enough to merit policies to stimulate population growth. Those of France, Germany, Japan, Italy, Greece, Poland and Hungary all range between 60 and 67.2 per cent. Iran is at 72.6 per cent, Brazil is at 78.7 per cent and Russia fares best at 83.7 per cent. All are lower than a suggested minimum threshold of 90 per cent indicating the support of reasonably buoyant biodiversity.

Policies to Reduce Population Size

Developing countries usually take the opposite approach to those in Europe, recognising the need to curb population growth. A few examples of policies in Africa, Asia and Latin America are considered below.

With more than 200 million people, Nigeria is the most densely populated country in Africa (Figure 10.4). In 2018, the country's finance minister called for a discussion about the country's birth rate, which is among the highest in the world. She pointed out that many families cannot even feed the children they have, not to mention access to good-quality healthcare or education, 'so it is necessary to talk about these things'. Perhaps doing more than talking would be a sound plan. In Kenya, the government passed a landmark policy in 2012 to manage its rapid population growth. The policy aimed to reduce the number of children a woman has over her lifetime from five in 2009 to three by 2030. The policy also included innovative targets for reducing child and maternal mortality, improved life expectancy and other reproductive health measures. This enlightened population control programme is clearer than that of neighbouring Tanzania, which was instituted in 2006. In this case, a wordy document aims to integrate economic activity and population growth, with a view to 'coordinating and influencing other policies, strategies and programmes that ensure sustainable development of the people and promoting gender equality and the empowerment of women'. This vague ambition reflects a lack of consensus in the Tanzanian government over the role of family planning in the National Population Policy. Even so, its implementation is primarily a policy for assisted family planning and therefore undoubtedly represents progress along sensible lines.

Figure 10.4. A vast crowd in densely populated Nigeria. *Source*: Michel Setboun/ Hulton Archive/Getty Images

Egypt has a substantial population problem. Its government has long implemented programmes to encourage smaller families and the use of birth control, but far more intrusive population control policies are on the cards. A two-child cap for welfare recipients and an incentive scheme for one-child families have been developed. Egyptian Prime Minister Mostafa Madbouly announced that everyone should realise that the runaway growth of Egypt's population is a big threat to the economic development in the country. Options included offering free school tuition throughout the entire course of their education to couples who have only one child and giving 'only' children priority when applying for a job on reaching employment age. One-child parents could also qualify for subsidised health insurance and free life insurance upon retirement.

On top of that, there could also be an honorary certificate for one-child parents, acknowledging their contribution to society by having fewer children and authorising couples to receive one-child-family government benefits. All in all, these somewhat draconian measures resemble a mixture of earlier Chinese incentives to reduce family size with the French approach of awarding medals, although in the latter case to increase fertility.

There are other countries in southern Asia as well as India with problems stemming from high population growth. Bangladesh is a prime example. It is one of the poorest and most densely populated countries in the world, with difficulties exacerbated by climate-driven sea-level rise along its fertile coastal plains. A policy instituted in 2004 included all the right things: it aimed to improve the status of family planning and maternal and child health including reproductive health service, and to improve the living standard through producing a balance between population and development. Total fertility rate reduction was to be achieved by raising awareness of family planning and increased use of planning methods. The net reproductive rate should be reduced such that the population should be stabilised by around 2060. This would require full availability and access to reproductive health services, especially family planning, for everyone and the provision of counselling services for adolescents. From a population of 137 million in 2004, the year of the policy declaration, numbers increased by 21 per cent to 166 million in 2021. Projections expect 193 million by 2050, after which a slow decline is expected, down to perhaps 191 million by the target turnaround date of 2060. So although the timescale hoped for in 2004 may well be met, the country will still have 54 million extra mouths to feed in mid-century, 40 per cent more than at the start of it. This looks like a truly formidable challenge.

Indonesia has also had to cope with increasing numbers of people, as outlined in Chapter 9, and with the high levels of environmental degradation and biodiversity losses that followed in their wake. The country's government became aware of the problem. and a family planning programme instituted in 1967 had population control in mind. A major accomplishment was bringing the contraceptive prevalence rate up to an outstanding 60 per cent, an achievement that halved the fertility rate from 5.2 to 2.6 per woman. Unfortunately, despite this progress, Indonesia's average population growth per year was still over 1.25 per cent for the decade ending in 2020. With over 270 million people in 2021, Indonesia ranked as the fourth most populous country in the world, and projections of 330 million by 2070 surpass the current population of the USA. Revitalisation of family planning has been criticised by religious groups who believe that it goes against their teachings. If the projections of an extra 60 million people come to pass, the prospects for both people and wildlife look bleak.

Then there is Latin America. Venezuela has no population policy but would undoubtedly benefit from one. The economic crisis that devastated the country in the 1980s, ending in a bailout by the International Monetary Fund in 1989, was exacerbated by population growth. Fertility remains above replacement level, contributing to a steady increase in population over time. In 2015, the population was six times the size that it was in 1950, and the UN projects that it will be nine times larger by 2050 and still growing at that point, despite a long-term decline in fertility. Fertility regulation has been largely absent among mature women in Colombia and Venezuela but is increasing among younger women. Strong family planning programmes in Colombia have resulted in a more rapid extension of contraceptive use, female sterilisation and the stopping after two children, relative to Venezuela. Clearly, this country could learn from its immediate neighbour.

With more than 14 million people, Guatemala is the most populous country in Central America. Its growth rate, at 2.5 per cent per year and driven mostly by fertility rather than immigration, is the highest in all of Latin America. It is projected that, by 2030, Guatemala's population will have grown from 17 million to 21.3 million people. Not only is overpopulation putting unprecedented pressure on natural resources, it is forcing increasingly more people into poverty. Fortunately, the total fertility rate is now declining, mostly due to increasing levels of women's education. Guatemala has a National Social Development and Population policy instigated in 2002, which prioritises a reduction in maternal and infant mortality as well as sexually transmitted diseases. Unfortunately, implementation of this policy has not yet been carried out as effectively as expected, but hopefully there will be increasing impacts in the coming years.

El Salvador is in many ways an unenviable place to live. It was considered the most violent country in the western hemisphere in 2015. With a population of approximately 6 million, it is the smallest but also the most densely populated country in Central America. El Salvador has one of the highest rates of teenage pregnancies in Latin America. The government has banned abortions, and those undergoing one face 5–8 years in jail. Even girls who miscarry can face criminalisation. There is no population policy, but there are hopeful signs of change. Women are having fewer children and more are using family planning, with particular increases in the use of injectable contraceptives. A 2008 survey found that fertility had dropped by 60 per cent in the previous 30 years, from 6.3 lifetime births per woman in the mid-1970s to 2.5 in the period between 2003 and 2008. A further reduction to replacement level was achieved by 2020, giving hope for a happier future in this stressed country.

BII estimates are in many cases higher in the selection of developing countries than in the developed ones. Lowest values, in the range of 53.4–67 per cent, were found in Bangladesh, El Salvador, Nigeria and Kenya. Indonesia, Tanzania and Guatemala had BII estimates between 70 and 77.6 per cent, while Colombia, Venezuela and Egypt ranged from 81.2 per cent to, in the case of Egypt, 93.5 per cent. Developing countries taking steps to curtail their population growth have often, so far, maintained a higher proportion of their biodiversity than developed countries seeking to stimulate it. The relationship between human population density and BII is complex and strongly influenced by country-specific factors. Some countries with low population densities, such as Australia, also have low BII estimates. Nevertheless, as shown in Chapter 9, BII estimates across the globe decline overall as a function of existing human population densities.

10.3 A Population Policy for the UK?

The Current Situation

There is no population policy in the UK and, at the time of writing, no immediate prospects for one, despite public opinion, as shown in Chapter 8, showing majority support for such a policy. It is notable that the Netherlands, with a high population density comparable to that of England, also has no official policy on the subject. At least, unlike many European countries, there are no attempts underway in either country to boost population growth, and again in both countries fertility is below replacement level. Immigration is the primary cause of population growth in the UK, and has been for many years. This presents an immediate problem within the UK because its component nations have divergent views on immigration policy. The Westminster government has taken an increasingly hard line in trying to reduce net immigration into the UK, but the Scottish government seeks to increase it (Scottish

Government, 2018) claiming on economic grounds that it is essential to the country's future prosperity. Whether or not this divergence remains a serious hindrance to progress depends in part on whether Scotland remains part of the UK in the medium- or long-term future.

Aspects of a UK Population Policy

What would be the primary objectives of a UK population policy, and how could these be achieved? As described in earlier chapters, the current British population size is already too high to sustain the country's wildlife on account, particularly, of expanding urban sprawl and the need for intensive farming to assure food supplies. The downside of having so many people extends way beyond wildlife issues and includes adverse impacts on human welfare and well-being. With human numbers in the UK approaching 70 million, how much smaller a number would constitute a desirable or feasible target? Unsurprisingly, this is both a complex and a controversial subject. It has been argued that on the basis of self-sufficiency in food and other vital resources the UK could support around 20 million people. However, this in turn depends on definitions of self-sufficiency and in particular on the standards of living implied by target numbers. In any case, such a drastic reduction of more than 70 per cent compared with current numbers is unachievable, short of some appalling catastrophe, within a time frame of at least decades. In practice, setting targets like this are unrealistic and essentially pointless. Far better would be to set in train policies that could at least reverse ongoing population increases and ideally initiate a slow but steady decline in the number of people in these crowded islands. The recent focus on climate change as the dominant environmental problem is understandable but is rarely considered in the perspective of population growth (Figure 10.5). Action to control global warming can be a distraction if it ignores a major underlying cause.

Aside from improving the prospects for wildlife and creating a better quality of life for people, there is also a moral imperative for reducing the UK's population size. The numbers of humans living in Britain, and in many other rich western countries, are not self-sustainable. They require the sequestration of unfairly large amounts of global resources, inevitably at the expense of poorer, developing nations. This situation could be resolved by two, not mutually exclusive, changes. Richer countries could reduce their standard of living, leaving more for others. This discrepancy has been highlighted for decades, but little has been done to redistribute

Figure 10.5. We need to talk. © Timothy Jay Newcomb

the world's wealth, largely because power lies with rich countries unwilling to make the necessary sacrifices. Another approach would be to gradually reduce human numbers in developed countries, leaving those people remaining to retain a high standard of living while making available the surplus resources to upgrade the lives of people in the developing world. Seen in this perspective, the parochial objectives of increasing populations in already wealthy nations is surely both selfish and immoral.

What, then, might population-reduction policies entail? We have certainly learnt what not to do. The coercive approaches of compulsory sterilisation in India and the enforced one-child policy in China are anathema to civilised democracies. Any attempt to develop a population policy in the UK will only succeed if it carries with it a consensus of the great majority of people living there. As described in Chapter 7, there are three key factors underpinning population growth: birth rate, death rate and immigration. How might these be influenced in favour of population reduction?

Death rates are perhaps the easiest factor to consider. These have declined continuously in many countries, leading to substantial increases in average longevity. This has contributed to the UK population increase, but almost certainly death rates will stabilise before long as natural ageing processes limit how far this trend can go. Longer lives have resulted from ongoing improvements in the standard of living, especially clean sanitation and effective medical interventions. Providing the necessary resources remain available, these benefits will presumably continue, but unless they include developments that retard intrinsic ageing mechanisms, they will surely arrest at some not far distant point. Of the three factors determining population growth, this is one that nobody would want to reverse. Long and healthy lives are an uncontentious ambition. The main worry here is that death rates might be taken out of our hands. High human population densities, especially in cities, are fertile ground for the spread of dangerous infections. As shown by the coronavirus 2019 (COVID-19) pandemic caused by severe acute respiratory syndrome coronavirus 2 (SARS-CoV-2), viruses spread all too rapidly in our interconnected world. As pathogens go, SARS-CoV-2 has proved relatively feeble, killing only a tiny fraction of infected communities. Even so, this has proved enough to cause millions of tragic deaths around the globe and serious suffering for very many more. A more dangerous microbe could cause catastrophic mortality, with no guarantee that even the best bioscience could halt its effects in reasonable time. SARS-CoV-2 targets the respiratory system, resulting in selectively high mortality in elderly people. There is no reason to expect that future pathogens will restrict their virulence to particular age groups. Lessening the possibilities of pathogen transfer from other animal species, the likely cause of not just COVID-19 but also of AIDS and Ebola (both of which cause disease across the age spectrum), is now a necessity. Reducing the human population size would diminish the possibilities for such high rates of spread, simultaneously reducing pressure on the bush meat and other wildlife trade. Unfortunately, most epidemiologists consider that, under current conditions, future pandemics are more or less inevitable.

Birth rates are often the primary cause of population increase. In much of the developing world, high fertility, with women typically bearing three or more children, has fuelled human expansion. In the UK, as outlined in Chapter 7, fertility has been below the replacement rate of just over two children per woman for several decades. In any country not suffering from population pressures, this might be good enough to obviate the need for an active population-reduction policy. Unfortunately, this is not the situation in the UK, where the existing population density is unsustainable without continuous and accelerating damage to the country's environment. High human numbers are causing ongoing problems for both people and wildlife. A good policy should encourage further reduction in fertility by incentives rather than by intimidation, and start with participatory persuasion. Governments in Europe have a poor record in coaxing people to change their family size, but this has mostly been directed towards promoting increases. Unsurprisingly, this has usually been resisted, requiring as it does an extra parenting effort beyond what individuals are otherwise inclined to opt for. A wide-ranging discussion explaining the problems resulting from overpopulation, specifically in the UK, should have a better chance of success, especially by emphasising that the choice will ultimately remain a personal one. Nevertheless, incentives are likely to be important. At the time of writing, parents can claim financial support from the government, although since 2017 this has been limited, in most cases, to just two children. This policy has been criticised on the basis that it penalizes the less well off, leaving those on higher incomes freer to have large families.

One way around this could be to reform income distribution and rejig the tax system in substantial ways. According to the ONS, income inequality in the UK, as well as that of total wealth, was continuously high over the decade preceding 2020. In that year, the income of the richest 20 per cent of people was over six times higher than that of the poorest 20 per cent, while the richest 10 per cent enjoyed 50 per cent more income than the poorest 40 per cent. Income inequality in the UK increased most recently because a benefits freeze hit the poorest households, reducing their incomes for two consecutive years. Thus, the median disposable income – that remaining after the deduction of taxes and social security charges – fell for the poorest fifth of the population by 4.3 per cent per year over the 2 years to April 2019. Britain has one of the highest levels of income inequality in the western world, and this has continued since the divide between rich and poor ballooned in the late 1970s and early 1980s. Redistribution of wealth in the UK is surely overdue. Ways of addressing this imbalance are well known in principle and include reductions in regressive taxes such as value-added tax, and increases in taxes for rich people and on the profits of wealthy corporations. This income could be used to increase the benefits of those on low wages, where employers are unable or unwilling to increase salaries to the point where comfortable living becomes the national norm. It would need to be substantially more than the so-called 'living wage', which is low enough to drive some people to a dependence on food banks. It should also suffice to support the raising of a couple of children without specific child benefits. But fairness about family size would need something more. A reasonable proposition is that those wealthy enough to raise more than two children should pay extra tax, on a per extra child basis, to compensate for the increased burden more people place on infrastructure and the environment. These changes would require a revolution in political will and would undoubtedly meet with sustained resistance. Nevertheless, if they or similar changes were implemented, there would be long-term benefits for the majority of British citizens.

As shown in Chapter 7, the third factor, net immigration, has been the primary driver of population growth in the UK since the 1990s. This has become a politically sensitive issue. Calls for constraints on immigration have been derided as racist because most newcomers are from poor countries and have predominantly non-white ethnic backgrounds. Racist connotations are inevitable in this situation, even when none are intended. Any control of incomers should – indeed must – apply to all countries, irrespective of origin. Serious attempts to reduce human numbers in the UK cannot ignore immigration as a major contribution to the situation. In this context, it will be vital to disconnect immigration control from the racism allegations that some political groups have continuously made. Immigrants have enriched the UK's cultural heritage from time immemorial, and have been a feature of the British nation at least since the Roman invasion 2,000 years ago. From the mid-twentieth century onwards, newcomers have been ever more ethnically divergent from white Anglo-Saxons, and many suffered shameful racist abuse during their settlement in Britain. Such ignorant attitudes have not gone away, but in general, a tolerant, multicultural society is gradually emerging. Most people now recognise the benefits that this has brought along the way, including everything from scientific skills to novel culinary delights. In 2001, Foreign Secretary Robin Cook called chicken tikka masala 'a true British national dish'. Cook was right. Curried chicken in a spicy sauce had become the UK's most popular meal. I and no doubt many others would concur with Cook's opinion that this is one of the many delights in a varied and progressive society.

Public opinion in a 2020 study was divided on the subject of immigration. About 44 per cent of people, the largest group, considered it was too high, while 39 per cent reckoned current levels were acceptable. Discontent is partly due to the very large numbers, hundreds of thousands per year, arriving legally in the UK and partly on account of the many hundreds seeking to arrive illegally by crossing the English Channel on ferries or in small boats. These people are often classified as economic migrants or asylum seekers, although the distinction between these categories is often unclear. For many years, the UK government has attempted to stem the flow of immigrants, thus far with very little success. What might a fair policy to achieve this end look like? Every year considerable numbers of people come to work in low-paid jobs that do not attract native labourers, especially on farms and in the hospitality industry. The NHS, too, has relied heavily on recruitment of newcomers. The government attitude seems to be that employers should pay enough to attract home-grown workers, but in the short term that does not look a likely prospect. Maybe a significant redistribution of wealth as proposed above would solve the problem, but at the moment that seems like pie in the sky. These workers, and also foreign students studying in the UK, make a significant contribution to immigrant numbers but most are temporary visitors. Sooner or later, most of them go home and do not make a permanent addition to the UK population. It is those immigrants aiming to start a new life in Britain and raise families here that have lasting impacts on population growth. In January 2021, the government introduced a points system in which potential immigrants are given a score and only those with the required skills are to be allowed in. Selection is therefore based on what types of workers the host country requires. Arguably, this by itself is too harsh, and a more enlightened points system might also take account of the needs of applicants, including dangers in their country of origin and the degree of poverty. In any case, serious attempt to address UK overpopulation certainly must include effective limits on future numbers of permanent immigrants. There is no escaping a need to do this because the existing British population is already too high to avoid long-term environmental degradation. According to the ONS, in 2020, 715,000 people immigrated into the UK while 403,000 left the country. A reduction of immigration numbers to less than those emigrating would therefore begin to reduce population pressure from immigration. On the basis of 2020 figures, and those of other recent years, allowing in tens of thousands of people each year would be compatible with a downward trend in the UK population size. This does not seem an unreasonable ambition.

10.4 Overview

Reading about the latest conservation success stories on the websites of NGOs such as those of the Wildlife Trusts and the RSPB is an uplifting experience. New nature reserves, thriving species responding to protection and management, successful reintroductions and the like are heart-warming developments for wildlife enthusiasts. The achievements of these dedicated practitioners are quite amazing and give hope for the future. Unfortunately, the big picture is very different from the impression so easily given by the good news. Overall, wildlife declines in the UK continue apace. Evidence accumulated over decades was summarised by Maclean (2010) and further documented by three State of Nature reports (Burns et al., 2013; Hayhow et al., 2016, 2019). Over this period, there has been no let-up in the downward trajectory of wildlife in Britain, leaving the country one of the most nature-depleted in the world. A BII estimate of less than 90 per cent indicates that the planetary boundary for biosphere integrity has been crossed. The average BII in the UK was 13th from the bottom of estimates from 236 countries and dependencies. Attempts to meet the UK biodiversity targets set by the government in 2010 have failed even more dismally than previous similar empty promises. Fourteen of 24 biodiversity indicators confirmed long-term decline including continued deterioration of UK habitats and species of European importance, as well as a decline in priority species, according to the 2021 UK biodiversity indicators report (DEFRA, 2021). The valiant efforts of so many dedicated conservationists are playing patch up, not catch up.

The proximal causes of wildlife declines in Britain have been well known for many years, and in this book I revisit them in the context of a general hypothesis that the primary cause is the impact of too many people competing for limited space and resources. In my view, this hypothesis is supported at least with respect to the two most important generators of wildlife declines, notably urbanisation and agricultural intensification. The correlations between amphibian and bird declines, two well-documented taxonomic groups, and the human population densities in Europe shown in Chapter 7 provide correlative support for the adverse impacts of human populations on wildlife. This connection is confirmed across the globe on the bases of BII estimates. But you will look in vain for anything other than passing comments on human numbers in the State of Nature reports, or on wildlife NGO websites and literature. Among multiple contributors to *Silent Summer* (Maclean, 2010), only Gaston and Evans specifically addressed the consequences of population growth, in their chapter on urbanisation and development. Without bringing human numbers into mainstream thinking in the context of wildlife and human futures, significant changes for the better look almost impossible. For far too long discussion about overpopulation in developed countries including the UK has been taboo in polite society. This needs to change. A humane population-reduction policy would not have a rapid beneficial effect but is vital for any chance of proper recovery for Britain's outstanding wildlife heritage in the longer term. To this end, it will be necessary to replace blinkered economic arguments that have consistently ignored the real, biological world in which human society functions. In the meantime, we can only applaud and support the ongoing efforts of dedicated conservation organisations to hold the fort in the hope that better times will one day come. Britain's countryside and wildlife heritage have inspired people for generations, but we must see clearly the scale of the challenge posed by overwhelming human numbers if our descendants are to inherit the delights we have so long enjoyed.

10.5 Population Matters

Population Matters is the go-to organisation for those concerned about damage being done by overpopulation, not just in the UK but all around the world. Originating as the Optimal Population Trust, what later became Population Matters was initiated by David Richardson and David Willey in the early 1990s. As of 2021, Population Matters had about 18,000 members. Prominent among those dedicated individuals running the charity have been Alistair Currie, John Guillebaud and Aubrey Manning. Many distinguished people have associated themselves with Population Matters as patrons, as listed in Chapter 8. One of these, David Attenborough, heads the Population Matters website (https://populationmatters.org/) with his famous quote: 'All our environmental problems become easier to solve with fewer people, and harder – and ultimately impossible – to solve with ever more people.' The Population Matters charity provides information on population facts, solutions to overpopulation problems, 'mythbusting' and an alarming ticking clock showing how the world's human numbers are increasing in real time. Population Matters spokespeople have campaigned relentlessly to get population issues onto the political agenda, efforts that are at last gradually bearing fruit. The impacts of overpopulation on wildlife and ecosystem degradation are among many other damaging consequences of population growth that are highlighted by Population Matters. It is certainly a charity worthy of, and in need of, public support. At present, it stands almost alone in striving for measures that are increasingly urgent if those who come after us are to enjoy a stable, rich and fulfilling environment.

References

Adams, W. M., Small, R. D. S. and Vickery, J. A. (2014). The impact of land use change on migrant birds in the Sahel. *Biodiversity*, 15, 101–108.

Akeroyd, J. (1999). *Plant Crime: Is the Law Working to Save Our Threatened Plants?* Plantlife. www.plantlife.org.uk/application/files/9114/8240/4595/Plant-crime-1999.pdf

Angold, P. G. (1997). The impact of a road upon adjacent heathland vegetation: effects on plant species composition. *Journal of Applied Ecology*, 34, 409–417.

Bacon, F. (1620). *Novum Organon*. Translated by Urbach, P. and Gibson, J. (1994). Chicago: Open Court.

Balfour-Browne, F. (1962). *Water Beetles and Other Things*. Dumfries, UK: Blacklock Farries & Sons.

Bartrip, P. W. J. (2008). Myxomatosis in 1950s Britain. *Twentieth Century British History*, 19, 83–105.

Beebee, T. J. C. (1995). Amphibian breeding and climate. *Nature*, 374, 219–220.

Beebee, T. J. C. (2013). Effects of road mortality and mitigation measures on amphibian populations. *Conservation Biology*, 27, 657–668.

Beebee, T. J. C. (2018). *Climate Change and British Wildlife*. London: Bloomsbury.

Beebee, T. J. C. and Griffiths, R. A. (2005). The amphibian decline crisis: a watershed for conservation biology? *Biological Conservation*, 125, 271–285.

Bell, T. (1837). *A History of British Quadrupeds, Including the Cetacea*. London: Van Voorst.

Bonnington, C., Gaston, K. J. and Evans, K. L. (2013). Fearing the feline: domestic cats reduce avian fecundity through trait-mediated indirect effects that increase nest predation by other species. *Journal of Applied Ecology*, 50, 15–24.

Bragg, R. and Atkins, G. (2016). *A Review of Nature-Based Interventions for Mental Health Care*. Commissioned Report No. 204. York, UK: Natural England. http://publications.naturalengland.org.uk/file/6567580331409408

British Association for Shooting & Conservation (2015). *Grouse Shooting and Management in the United Kingdom: Its Value and Role in the Provision of Ecosystem Services*. Rossett, UK: British Association for Shooting & Conservation. https://basc.org.uk/download/67954/

British Sea Fishing (2016). Commercial fishing. Online article. https://britishseafishing.co.uk/commercial-fishing/

Burns, F., Eaton, M. A., Gregory, R. D. et al. (2013). *State of Nature Report*. The State of Nature Partnership. www.rspb.org.uk/images/stateofnature_tcm9-345839.pdf

Burns, F., Eaton, M. A., Barlow, K. E. et al. (2016). Agricultural management and climatic change are the major drivers of biodiversity change in the UK. *PLoS One*, 11, e0151595.

Burns, F., Eaton, M. A., Hayhow, D. B. et al. (2018). An assessment of the state of nature in the United Kingdom: a review of findings, methods and impact. *Ecological Indicators* 94, 226–236.

Burton, L. R. (2003). The Mersey Basin: an historical assessment of water quality from an anecdotal perspective. *Science of the Total Environment*, 314–316, 53–66.

Carson, R. (1962). *Silent Spring*. Boston, MA: Houghton Mifflin Harcourt.

Chantrey, J., Dale, T. D., Read, J. M. et al. (2014). European red squirrel population dynamics driven by squirrelpox at a gray squirrel invasion interface. *Ecology and Evolution*, 4, 3788–3799.

Clegg, J. (1956). *The Observer's Book of Pond Life*. London: Frederic Warne & Co.

Conrad, K. F., Woiwod, I. P., Parsons, M., Fox, R. and Warren, M. S.(2004). Long-term population trends in widespread British moths. *Journal of Insect Conservation*, 8, 119–136.

Cooke, A. S. (1972). Indications of recent changes in status in the British Isles of the frog (*Rana temporaria*) and the toad (*Bufo bufo*). *Journal of Zoology*, 167, 161–178.

Cooke, A. S. (2011). The role of road traffic in the near extinction of Common Toads (*Bufo bufo*) in Ramsey and Bury. *Nature in Cambridgeshire*, 53, 45–50.

Cooke, S. C., Balmford, A., Johnston, A., Newson, S. E. and Donald, P. F. (2020). Variation in abundances of common bird species associated with roads. *Journal of Applied Ecology*, 57, 1271–1282.

Coombes, E. G., Jones, A. P. and Sutherland, W. J. (2008). The biodiversity implications of changes in coastal tourism due to climate change. *Environmental Conservation*, 35, 319–330.

Corbett, K. (1989). *Conservation of European Reptiles and Amphibians*. London: Christopher Helm.

Dadam, D., Robinson, R. A., Clements, A. et al. (2019). Avian malaria-mediated population decline of a widespread iconic bird species. *Royal Society Open Science*, 6, 182197.

Dasgupta, P. (2021). *The Economics of Biodiversity: The Dasgupta Review*. London: HM Treasury. www.gov.uk/government/publications/final-report-the-economics-of-biodiversity-the-dasgupta-review

Davies, Z. G., Fuller, R. A., Lorama, A. et al. (2009). A national scale inventory of resource provision for biodiversity within domestic gardens. *Biological Conservation*, 142, 761–771.

DEFRA (2021). *UK Biodiversity Indicators 2021 Revised*. London: DEFRA. https://assets.publishing.service.gov.uk/government/uploads/system/uploads/attachment_data/file/926506/UKBI-2020-A.pdf

Dennis, E. B., Morgan, B. J. T., Roy, D. B. and Brereton, T. M. (2017). Urban indicators for UK butterflies. *Ecological Indicators*, 76, 184–193.

Department for Transport (2019). *National Travel Survey: England 2018*. London: Department for Transport. https://assets.publishing.service.gov.uk/government/uploads/system/uploads/attachment_data/file/823068/national-travel-survey-2018.pdf

Ehrlich, P. (1968). *The Population Bomb*. New York, NY: Ballantine Books.

Ehrlich, P. R. and Ehrlich, A. H. (1996). *The Betrayal of Science and Reason*. Washington, DC: Island Press.

Eiseley, L. (1961). *The Man Who Saw Through Time*. New York, NY: Charles Scribner's Sons.

Eiseley, L. (1979). *Darwin and the Mysterious Mr. X*. New York, NY: E. P. Dutton,.

Estes, A. B., Kuemmerle, T., Kushnir, H., Radeloff, V. C. and Shugart, H. H. (2012). Land-cover change and human population trends in the greater Serengeti ecosystem from 1984–2003. *Biological Conservation*, 147, 255–263.

European Environment Agency (2020). *State of Nature in the EU*. Copenhagen: European Environment Agency. www.eea.europa.eu/publications/state-of-nature-in-the-eu-2020/at_download/file

Evans, M. R. (2019). Will natural resistance result in populations of ash trees remaining in British woodlands after a century of ash dieback disease? *Royal Society Open Science*, 6, 190908.

FAO (2016). *The State of World Fisheries and Aquaculture 2016: Contributing to Food Security and Nutrition for All*. Rome: FAO. www.fao.org/3/a-i5555e.pdf

Fisher, M. C., Garner, T. W. J. and Walker, S. F. (2009). Global emergence of *Batrachochytrium dendrobatidis* and amphibian chytridiomycosis in space, time, and host. *Annual Review of Microbiology*, 63, 291–310.

Ford, E. B. (1945). *Butterflies*. London: Collins.

Foreman, D. (2011). *Man Swarm, and the Killing of Wildlife*. Durango, CO: Raven's Eye Press.

Furness, E. N. and Robinson, R. A. (2019). Long-term declines in winter body mass of tits throughout Britain and Ireland correlate with climate change. *Ecology and Evolution*, 9, 1202–1210.

Gadow, H. (1909). *Amphibia and Reptiles*. London: MacMillan and Co.

Gardner, E., Julian, A., Monk, C. and Baker, J. (2019). Make the Adder Count: population trends from a citizen science survey of UK adders. *Herpetological Journal*, 29, 57–70.

Gilburn, A. S., Bunnefeld, N., Wilson, J. M. et al. (2015) Are neonicotinoid insecticides driving declines of widespread butterflies? *PeerJ*, 3, e1402.

Gillings, S. (2019). Bird responses to housing development in intensively managed agricultural landscapes. *Urban Ecosystems*, 22, 1007–1017.

Glover, J. (2019). *Landscapes Review: National Parks and AONBs*. London: DEFRA. www.gov.uk/government/publications/designated-landscapes-national-parks-and-aonbs-2018-review

Gotmark, F. and Andersson, M. (2020). Human fertility in relation to education, economy, religion, contraception, and family planning programs. *BMC Public Health*, 20, 265.

Goulson, D. (2013). *A Sting in the Tail*. London: Jonathan Cape.

Goulson, D. (2017). Are neonicotinoids killing birds, if so where are they? Online article. Nurturing Nature. https://nurturing-nature.co.uk/bumblebees-and-their-ecology/are-neonicotinoids-killing-birds-if-so-where-are-they/

Goulson, D. and Nicholls, E. (2016). The canary in the coalmine; bee declines as an indicator of environmental health. *Science Progress*, 99, 312–326.

GOV.UK (2017). *Local Planning Authority Green Belt: England 2016/17*. London: HM Government. https://assets.publishing.service.gov.uk/government/

uploads/system/uploads/attachment_data/file/642684/Green_Belt_Statistics_England_2016-17.pdf).

GOV.UK (2020). *New National Parks and Thousands of Green Jobs Under Plans to Build Back Greener.* Press release. www.gov.uk/government/news/new-national-parks-and-thousands-of-green-jobs-under-plans-to-build-back-greener

Gregus, J. and Guillebaud, J. (2020). Doctors and overpopulation 48 years later: a second notice. *European Journal of Contraception & Reproductive Health Care*, 25, 409–416.

Hallmann, C. A., Sorg, M., Jongejans, E. et al. (2017) More than 75 percent decline over 27 years in total flying insect biomass in protected areas. *PLoS One*, **12**, e0185809.

Harfoot, M., Glaser, S. A. M., Tittensor, D. P. et al. (2018). Unveiling the patterns and trends in 40 years of global trade in CITES-listed wildlife. *Biological Conservation*, 223, 47–57.

Harrison, M. (2016). *Rain: Four Walks in English Weather.* London: Faber & Faber.

Hart, J. D., Milsom, T. P., Fisher, G. et al. (2006). The relationship between yellowhammer breeding performance, arthropod abundance and insecticide applications on arable farmland. *Journal of Applied Ecology*, 43, 81–91.

Hayhow, D. B., Burns, F., Eaton, M. A. *et al.* (2016). *State of Nature 2016.* The State of Nature Partnership. www.rspb.org.uk/globalassets/downloads/documents/conservation-projects/state-of-nature/state-of-nature-uk-report-2016.pdf

Hayhow, D. B., Eaton, M. A., Stanbury, A. J. et al. (2019). *State of Nature 2019.* The State of Nature Partnership. https://rspb.org.uk/our-work/state-of-nature-report/

Hinsley, A., de Boer, H. J., Fay, M. F. et al. (2018). A review of the trade in orchids and its implications for conservation. *Botanical Journal of the Linnean Society*, 186, 435–455.

Hodgson, R. G. (2020). *Overcrowded Islands? The Challenges of Demographic Change for the United Kingdom.* London: Civitas. www.civitas.org.uk/content/files/Overcrowded-Islands1.pdf

Holderegger, R. and Di Giulio, M. (2010). The genetic effects of roads: a review of empirical evidence. *Basic and Applied Ecology*, 11, 522–531.

Hopkins, J. (2018). Something in the air, soil and water: nitrogen, phosphorus and British wildlife. *British Wildlife*, 29, 273–279.

Hopkins, J. (2019). The insect decline – time to limit pesticide use? *British Wildlife*, 31, 40–45.

Hudson, W. H. (1900). *Nature in Downland.* London: J. M. Dent and Sons.

ICES (2020). Greater North Sea ecoregion – fisheries overview, including mixed-fisheries considerations. In *Report of the ICES Advisory Committee, 2020.* ICES Advice 2020. https://doi.org/10.17895/ices.advice.7605

Inman, A., Brooker, E., Dolman, S., McCann, R. and Wilson, M. W. (2016). The use of marine wildlife-watching codes and their role in managing activities within marine protected areas in Scotland. *Ocean and Coastal Management*, 132, 132–142.

Jackson, S. F. and Gaston, K. J. (2008). The unpredictability of favourability: condition assessment and protected areas in England. *Biological Conservation*, 17, 749–764.

Jarvis, L. E., Hartup, M. and Petrovan, S. O. (2019). Road mitigation using tunnels and fences promotes site connectivity and population expansion for a protected amphibian. *European Journal of Wildlife Research*, 65, 27.

Jefferies, R. (1879). *Wildlife in a Southern County*. London: Thomas Nelson and Son.

Jiguet, F., Sunnen, L., Prévot, A. C. and Prince, K. (2019). Urban pigeons losing toes due to human activities. *Biological Conservation*, 240, 108241.

Kettel, E. F., Gentle, L. K., Yarnell, R. W. and Quinn, J. L. (2019). Breeding performance of an apex predator, the peregrine falcon, across urban and rural landscapes. *Urban Ecosystems*, 22, 117–125.

Kull, T. and Hutchings, M. J. (2006). A comparative analysis of decline in the distribution ranges of orchid species in Estonia and the United Kingdom. *Biological Conservation*, 129, 31–39.

Laville, S. (2020). Air pollution a cause in girl's death, coroner rules in landmark case. The *Guardian* online article. www.theguardian.com/environment/2020/dec/16/girls-death-contributed-to-by-air-pollution-coroner-rules-in-landmark-case

Lawton J. H., Brotherton, P. N. M., Brown, V. K., et al. (2010). 'Making space for nature': a review of England's wildlife sites and Ecological Network. Press release. www.gov.uk/government/news/making-space-for-nature-a-review-of-englands-wildlife-sites-published-today

Lewis, S. A. (2021). Deer culling in Britain: what's the problem and why are deer culled? Countryfile Magazine. www.countryfile.com/wildlife/mammals/deer-culling-in-britain-facts-and-statistics/.

Lister, A. (2014). *Mammoths: Ice Age Giants*. London: Natural History Museum.

Lowe, A., Rogers, A. C. and Durrant, K. L. (2014). Effect of human disturbance on long-term habitat use and breeding success of the European Nightjar, *Caprimulgus europaeus*. *Avian Conservation and Ecology*, 9, 6.

Lutz, W. and Qiang, R. (2002). Determinants of human population growth. *Philosophical Transactions of the Royal Society London B*, 357, 1197–1210.

Mabey, R. (2010). *A Brush with Nature*. London: BBC Books.

Maclean, N. (ed.) (2010). *Silent Summer*. Cambridge: Cambridge University Press.

Mader, H. J. (1984). Animal habitat isolation by roads and agricultural fields. *Biological Conservation*, 29, 81–96.

Malthus, T. R. (1798). *An Essay on the Principle of Population, as it Affects the Future Improvement of Society with remarks on the Speculations of Mr. Godwin, M. Condorcet, and Other Writers*. London: J. Johnson.

Marn, N., Jusup, M., Kooijman, S. A. L. M. and Klanjscek, T. (2020). Quantifying impacts of plastic debris on marine wildlife identifies ecological breakpoints. *Ecology Letters*, 23, 1479–1487.

Marren, P. (2005). *The New Naturalists*, 2nd ed. London: HarperCollins.

McCarthy, M. (2015). *The Moth Snowstorm*. London: John Murray.

Meadows, D. H., Anders, J., Meadows, D. L. and Behrens, W. W. (1972). *The Limits to Growth: A Report for the Club of Rome's Project on the Predicament of Mankind*. New York, NY: Universe Books.

Mikola, J., Miettinen, M., Lehikoinen, E. and Lehtil, K. (1994). The effects of disturbance caused by boating on survival and behavior of velvet scoter *Melanitta fusca* ducklings. *Biological Conservation*, 67, 119–124.

Milton, N. (1994). Comment – nature conservation and arable farming. *British Wildlife*, 5, 229–235.

Morrissey, C. A., Stanton, D. W. G., Pereira, M. J. et al. (2013). Eurasian dipper eggs indicate elevated organohalogenated contaminants in urban rivers. *Environmental Science & Technology*, 47, 8931–8939.

Natural England (2015). *Green Bridges. A Literature Review*. Commissioned Report NECR181. http://publications.naturalengland.org.uk/file/6296975990325248

Newbold, T., Hudson, L. N., Arnell, A. P. et al (2016). Has land use pushed terrestrial biodiversity beyond the planetary boundary? A global assessment. *Science*, 353, 288–291.

Newton, I. (2018). Seeds and seed-eating birds: casualties of agricultural change. *British Wildlife*, 29, 177–183.

Nunez, S., Arets, E., Alkemande, R., Verwer, C. and Leemans, R. (2019). Assessing the impacts of climate change on biodiversity: is below 2 °C enough? *Climate Change*, 154, 351–365.

Office for National Statistics (2018). UK natural capital: developing semi-natural grassland ecosystem accounts. London: Office for National Statistics. www.ons.gov.uk/economy/environmentalaccounts/methodologies/uknaturalcapitaldevelopingseminaturalgrasslandecosystemaccounts

Ormerod, S. J. and Durance, I (2009). Restoration and recovery from acidification in upland Welsh streams over 25 years. *Journal of Applied Ecology*, 46, 164–174.

Orros, M. E. and Fellowes, M. D. E. (2015). Wild bird feeding in an urban area: intensity, economics and numbers of individuals supported. *Acta Ornithologica*, 50, 43–58.

Pennant, T. (1776). *British Zoology*. London: Benjamin White.

Peterken, G. (2013). *Meadows*. Oxford, UK: British Wildlife Publishing.

Petrenko, C., Paltseva, J. and Searle, S. (2016). *Ecological Impacts of Palm Oil Expansion in Indonesia*. White Paper. Washington, DC: International Council on Clean Transportation. https://theicct.org/wp-content/uploads/2021/06/Indonesia-palm-oil-expansion_ICCT_july2016.pdf

Petrovan, S. O. and Schmidt, B. R. (2016). Volunteer conservation action data reveals large- scale and long-term negative population trends of a widespread amphibian, the common toad (Bufo bufo). *PLoS One*, **11**, e0161943.

Pickard-Cambridge, O. (1893–94). Reptiles of Dorset. *Proceedings of Dorset Natural History Field Club*, 15, 90–102.

Piersma, T. (2014). *Guests of Summer*. Thetford, UK: BTO Books.

Potter, C., Harwood, T., Knight, J. and Tomlinson I. (2011). Learning from history, predicting the future: the UK Dutch elm disease outbreak in relation to contemporary tree disease threats. *Philosophical Transactions of the Royal Society B*, 366, 1966–1974.

Price, S. J., Garner, T. W. J., Cunningham, A. A., Langton, T. E. S. and Nichols, R. A. (2016). Reconstructing the emergence of a lethal infectious disease of wildlife supports a key role for spread through translocations by humans. *Proceedings of the Royal Society B*, 83, 20160952.

Ray, J. (1686) *Historia Plantarum*. London: Clark.

Remacha, C., Delgado, J. A., Bulaic, M. and Pérez-Tris, J. (2016). Human disturbance during early life impairs nestling growth in birds inhabiting a nature recreation area. *PLoS One*, **11**, e0166748.

Riggio, J., Jacobson, A., Dollar, L., et al. (2013). The size of savannah Africa: a lion's (Panthera leo) view. *Biodiversity and Conservation*, 22, 17–35.

Ripple, W. J., Wolf, C., Newsome, T. M. et al. (2017) World scientists' warning to humanity: a second notice. *BioScience*, 67, 1026–1028.

Ripple, W. J., Wolf, C., Newsome, T. M., Barnard, P. and Moomaw, W. R. (2020). The climate emergency, forests, and transformative change. *BioScience*, 70, 8–12.

Robinson, M. (1990). *Impact of Improved Land Drainage on River Flows*. Report no. 113. Wallingford, UK: Institute of Hydrology. http://nora.nerc.ac.uk/id/eprint/7349/1/IH_113.pdf

Rodway-Dyer, S. J. and Walling, D. E. (2010). The use of ^{137}Cs to establish longer-term soil erosion rates on footpaths in the UK. *Journal of Environmental Management*, 91, 1952–1962.

Scottish Government (2018). *Scotland's Population Needs and Migration Policy*. Edinburgh, UK: Scottish Government. www.gov.scot/publications/scotlands-population-needs-migration-policy/

Shaw, C. (2007). Fifty years of United Kingdom national population projections: how accurate have they been? *Population Trends*, 128, 8–23.

Shoard, M. (1980). *The Theft of the Countryside*. Isleworth, UK: Maurice Temple Smith.

Shragg, K. I. (2015). *Move Upstream*. Minneapolis-St Paul, MN: Freethought House.

Slater, F. (1994). Wildlife road casualties. *British Wildlife*, 5, 214–221.

Smith, H. (2020). The fen raft spider-from uncertain past to uncertain future. *British Wildlife*, 32, 98–109.

Smith, K. M., Zambrana-Torrelio., White, A., Amussen, M., Machalaba, C. et al. (2017). Summarizing US wildlife trade with an eye toward assessing the risk of infectious disease. *EcoHealth*, 14, 29–39.

Smith, M. A. (1951). *British Amphibians and Reptiles*. London: Collins.

Smith, P. H. (1999). *The Sands of Time. An Introduction to the Sand Dunes of the Sefton Coast*. Liverpool, UK: National Museums and Galleries on Merseyside.

Stallings, C. D. (2009). Fishery-independent data reveal negative effect of human population density on caribbean predatory fish communities. *PLoS One* **4**, e5333.

Stein-Bachinger, K., Gottwald, F., Haub, A. and Schmidt, E. (2021). To what extent does organic farming promote species richness and abundance in temperate climates? A review. *Organic Agriculture*, 11, 1–12.

Stenning, M. (2018). *The Blue Tit*. London: Bloomsbury (Poyser).

Sterry, P. (2006). *Complete British Wild Flowers*. London: HarperCollins.

Tansley, A. G. (1946). *Our Heritage of Wild Nature*. Cambridge, UK: Cambridge University Press.

Tett, P., Benjamins, S., Black, K. D. et al. (2018). *Review of the Environmental Impacts of Salmon Farming in Scotland. Executive Summary and Main Report. Issue 01, January 2018*. Edinburgh, UK: The Scottish Parliament. www.parliament.

scot/S5_Environment/General%20Documents/20180125_SAMS_Review_of_Environmental_Impact_of_Salmon_Farming_-_Report.pdf

Thomas, C. D. (2018). *Inheritors of the Earth – How Nature Is Thriving in an Age of Extinction*. London: Penguin Books.

Thompson, K. and Jones, A. (1999). Human population density and prediction of local plant extinction in Britain. *Conservation Biology*, 13, 185–189.

Tree, I. (2018). *Wilding*. London: Pan Macmillan..

Tuyttens, F. A. M., Delayah, T. R. J., Macdonald, D. W., et al. (2000). Spatial perturbation caused by a badger (*Meles meles*) culling operation: implications for the function of territoriality and the control of bovine tuberculosis (*Mycobacterium bovis*). *Journal of Animal Ecology*, 69, 815–828.

Uberoi, E., Hutton, G., Ward, M. and Ares, E. (2021). *UK Fisheries Statistics*. London: House of Commons Library. https://researchbriefings.files.parliament.uk/documents/SN02788/SN02788.pdf

UNEP Global Environmental Alert Service (2012). *One Planet, How Many People? A Review of Earth's Carrying Capacity*. Nairobi, Kenya: UN Environment Programme. https://na.unep.net/geas/archive/pdfs/geas_jun_12_carrying_capacity.pdf

UN Population Information Network (1974). *World Population Plan of Action*. New York, NY: United Nations. www.un.org/en/development/desa/population/migration/generalassembly/docs/globalcompact/E_CONF.60_19_Plan.pdf

US Department of State (2010). *Population*. Washington, DC: US Department of State. www.state.gov/other-policy-issues/population/

White, G. (1789). *The Natural History and Antiquities of Selborne*. London: Benjamin White (but with some 300 later editions).

Williams, B. M., Baker,P. J., Thomas, E., et al. (2018). Reduced occupancy of hedgehogs (*Erinaceus europaeus*) in rural England and Wales: the influence of habitat and an asymmetric intra-guild predator. *Scientific Reports*, 8, 12156.

Williams, F., Eschen, R., Harris, A., et al. (2010). *The Economic Cost of Invasive Non-Native Species on Great Britain*. Wallingford, UK: CAB International.

Wilson, P. J. (1992).Britain's arable weeds. *British Wildlife*, 3, 149–161.

World Wide Fund for Nature (2020). *Living Planet Report 2020*. Gland, Switzerland: WWF. https://livingplanet.panda.org/en-gb/

Wright, P. L., Zhang, L., Cheng, I., Aherne, J. and Wentworth, G. R. (2018). Impacts and effects indicators of atmospheric deposition of major pollutants to various ecosystems – a review. *Aerosol and Air Quality Research*, 18, 1953–1992.

Wynes, S. and Nicholas, K. A. (2017). The climate mitigation gap: education and government recommendations miss the most effective individual actions. *Environmental Research Letters* **12**, 074024.

Yurday, E. (2021). UK food prices 2020. NimbleFins online article. www.nimblefins.co.uk/uk-food-prices

Zayed, Y. and Loft, P. (2019). *Agriculture: Historical Statistics*. House of Commons Briefing Paper No. 3339. London: House of Commons Library. https://researchbriefings.files.parliament.uk/documents/SN03339/SN03339.pdf

Zick, D., Gassner, H., Filzmoser, P., et al. (2006). Changes in the fish species composition of all Austrian lakes >50 ha during the last 150 years. *Fisheries Management and Ecology*, 13, 103–111.

Index